£2.95

SOLIDS: AN INTRODUCTION

Spots.

Solids:

an introduction

A. HART-DAVIS, D.Phil., M.A.

McGraw-Hill

London . New York . St Louis . San Francisco . Düsseldorf . Johannesburg
Kuala Lumpur . Mexico . Montreal . New Delhi . Panama . Paris . São Paulo
Singapore . Sydney . Toronto

Published by
McGraw-Hill Book Company (UK) Limited
Maidenhead . Berkshire . England

Library of Congress Cataloging in Publication Data

Hart-Davis, A
Solids : an introduction

Includes bibliographical references.
1. Solids. I. Title.
QC176.H28 530.4'1 75-1169
ISBN 0-07-084058-X

Printed and bound in Great Britain

Preface

This book is intended for undergraduates embarking on an introductory course in solid state physics. It does not attempt to cover the whole subject up to degree level: final-year students will need to use a weightier book. I hope that this book will help those who find that a large and very learned text on a subject new to them can be rather frightening. The reader is assumed to be familiar with some of the basic principles of statistical mechanics and of wave mechanics.

In writing this book I have limited myself to perfect or very-nearly-perfect crystals and have made no attempt to deal with large dislocations. However, many important physical properties of real solids depend strongly on imperfections. The student who wishes to broaden his knowledge of non-perfect solids can find non-mathematical treatments in, for example, G. C. E. Olds *Metals and Ceramics* Oliver and Boyd (1968) or *Materials* (A Scientific American Book) W. H. Freeman (1967). A more rigorous treatment of dislocations such as will be found in the standard texts is, I feel, a little hard to understand unless one is familiar with the properties of perfect crystals and is perhaps better left until a later course.

I have used SI units throughout, except for a few such as ångströms and electronvolts which are occasionally more convenient. The formulae are all in SI including those using electrical and magnetic quantities. I have referred to B, the force vector, as the magnetic *field*. This nomenclature, with B the field and H the magnetic intensity, is not the one recommended by the Royal Society but it is adopted by many authors and seems to me more sensible. The units of B are tesla whether you wish to think of it as the magnetic field or as the induction.

All the worked examples are parts of questions set for the Oxford University Physics Finals Examinations and I am grateful to the Clarendon Press for permission to use them. The Universities of Bristol and Manchester have kindly allowed me to make use of questions set in their examinations. These are not all finals problems. I should also like to thank Cornell University Press for permission to use some of the data that appears in Tables 3.2 and 3.3. I wish to thank my husband for his patience and for help with various bits of chemistry. Finally, I must acknowledge the help of Dr J. A. D. Matthews of the University of York who read the typescript very carefully and made many suggestions for improvements. I have not always carried out his suggestions and any remaining errors in facts or in presentation are mine alone.

A. Hart-Davis

Contents

1. Crystal Structure

Introduction

This is a book about some of the basic concepts that are needed for the study of solids. Discussion will be limited to simple crystalline solids. A crystalline solid is one in which the basic units (atoms, ions, or molecules) are arranged in a lattice – a repetitive three-dimensional framework. By simple I mean that the basic units should be simple, and this definition therefore excludes crystals composed of large organic molecules. Which solids can be called simple crystals? Let us consider some different classes of solids.

Glass is an *amorphous* solid. This means that although there is some short-range order within the solid there is no long-range order as in crystals. In fact glass is more like a liquid in microscopic structure than it is like a solid and is often thought of as a super-cooled liquid.

Next we may consider mineral salts such as sodium chloride. These do indeed form simple crystals, which are often large enough to be seen clearly with the naked eye. The symmetric shape of many crystals is an indication of some underlying order on a microscopic scale.

'Plastics' are synthetic polymers, and fibres such as cotton and wool are examples of natural polymers. Polymers are excluded from our studies because a polymer has, by definition, huge molecules. Some polymers do show various degrees of crystallinity.

What about ceramics? The word 'ceramic' is not very clearly defined but can be taken roughly to mean a substance which is refractory, i.e., resistant to heat and corrosion. It consists of one or more metals (or an element such as silicon) in combination with a non-metallic element – usually oxygen. On hearing the word 'ceramic' most people immediately think of pottery. Such a ceramic will have a very complicated microstructure, consisting perhaps of tiny crystals of aluminium silicate held together by a silica glass. However, some ceramics are simple crystalline solids, for example magnesium oxide MgO, alumina Al_2O_3, and aluminium silicate itself.

Metals also form simple crystals. It should be emphasized that metals do not normally occur as single crystals. A metal is made up of many tiny crystals or grains (of the order of 10^{-5} m) arranged in random orientation. (Actually no crystal, metal or non-metal, even when free from impurities, can be truly perfect at any finite temperature, as we shall see in chapter 3.)

There are yet other types of solids which we shall come across later that are simple crystals. Perhaps it would be better to modify the second sentence of this chapter and to say that discussion will be limited to solids which can, in principle, form perfect single crystals built up from simple units. The first chapter is devoted entirely to the study of the perfect lattice. Information as to how a real crystal will differ from this theoretical ideal and on properties (many of them of great practical value) which are due to the imperfections will be found in later chapters.

Binding in crystals

In order that a substance shall exist as a crystal there must be forces between the atoms, ions, or molecules which are strong enough to hold the lattice together against the disruptive forces of thermal motion. If the temperature is raised sufficiently the lattice structure will break down. The stronger the bonds, the higher the temperature at which this breakdown occurs. We must now think about the different types of bonds which can exist between the basic units of a crystal. But first it is necessary to study briefly the structure of the atom.

Each element in the periodic table is distinguished from all the others by a characteristic value of the positive charge on the nucleus of an atom of this element. The magnitude of this charge (in units of the charge on the proton) is an integer known as the atomic number of the element. When an atom is in a neutral state, there must be a number of electrons, equal to the atomic number, in orbits around the nucleus since each electron has a negative charge of the same magnitude as the positive charge on the proton. The possible electron states around the nucleus have quantized energy levels and these states are grouped into 'shells'. A closed shell of electrons has no net angular momentum and no magnetic moment. The allowed states occur in pairs – one state corresponds to electron spin 'up' and the other to electron spin 'down' – and we can think of each electron as being in one of these allowed states. (This is not strictly true, for the electrons are indistinguishable from one another.) Consider what will happen if we put electrons around a nucleus to build up an atom in its ground state (lowest possible energy state). The first electron goes into one of the two allowed states with the lowest energy. The next goes into the other state of lowest energy with its spin anti-parallel to that of the first electron. The third electron occupies one of the next-lowest-energy states, etc. If the atomic number is such that the number of electrons needed to produce a neutral atom will be just sufficient to fill one or more shells, leaving other shells empty, then we have a noble-gas element. (See also chapter 6.) The noble gases are chemically very unreactive.

Suppose the atomic number is such that we do not get a closed-shell configuration in the neutral atom. The electrons outside closed shells are called the valence electrons and the number of these will determine the way that the atom behaves chemically and the type of bond that it will form with other atoms.

When there are only one, or two, electrons outside a closed shell, the atom will be prepared to give these up in order to obtain a stable closed-shell configuration which all atoms would like to possess. Of course, removal of electrons leaves a positively charged *ion* and therefore some energy is needed for this process. The greater the atomic number the more loosely bound is the extra electron and the smaller the amount of energy needed to ionize the atom. If there were many electrons outside a closed shell the energy involved would be too great, and some other mechanism for obtaining the closed shell would be preferred. Atoms that give up their outer electrons easily are called metals and we are all familiar with the more common ones.

Some atoms, the non-metals, have as it were the opposite problem. Instead of having one or two electrons outside a closed shell, they need only one or two electrons to complete the shell. In this situation, one way for the atom to gain the closed-shell configuration is for it to gain the extra electron or electrons and form a negative ion. Examples of non-metals are the halogens (chlorine, bromine, etc.) which need just one extra electron to complete their outer shell and to form the Cl^-, Br^-, etc., ions. All the halogens actually give out energy in the process. Oxygen and sulphur are two non-metals which need to gain two electrons and they can therefore form the ions O^{2-} and S^{2-}. However, they need energy to do this and so tend to form bonds which are partly ionic and partly covalent.

Other elements have shells about half filled. Too much energy would be required to form either a negative or a positive ion with a noble-gas electronic structure. Such atoms compromise and share electrons with other atoms that are in a similar predicament. They tend to form covalent bonds.

We are now in a position to study the bonding mechanisms in crystals. The main types of bonding are as follows:

1. The metallic bond. As the name suggests, this form of bonding is typical of metallic crystals. A metallic lattice can be visualized as consisting of positive metal ions bound together by an electron glue. Each metal atom donates its valence electron (or electrons) to this glue and becomes positively charged. Metallic bonds are comparatively weak – or rather the bond between any two metal atoms is weak – and the atoms occupy more space than one might expect (Table 1.1) from just knowing the size of the metal ions as they occur in ionic crystals (see later). However, the cohesive energy of a metal atom in a crystal is fairly large because, although each single bond is weak, an atom has an interaction not only with its nearest neighbours but with atoms much more distant in the lattice. The sum of all these bond energies is quite large. For a metal to have a large cohesive energy it must have as many neighbours as possible (with the restriction on nearest-neighbour spacing) and will therefore crystallize in one of the close-packed structures (see p. 18). The actual metallic bonds are not strongly directional and so the layers of atoms in the crystal will slide over one another quite easily. This explains why metals are ductile and malleable rather than brittle.

TABLE 1.1

	Sodium	Silver	Iron
Ionic radius	0.9Å	1.2	0.8 Fe^{2+}
Atomic radius in metal	1.9	1.4	1.2
Cohesive energy in metal lattice	1.1 eV	3.0 eV	4.3 eV

Table 1.1 gives a few interesting facts about sodium, iron, and silver. The cohesive energy is the energy gained (per atom) when stationary atoms an infinite distance apart are brought together into a metal crystal. It is the potential energy of an atom at infinity, minus the potential energy in the lattice, minus the zero-point vibrational energy (see chapter 2).

2. The ionic bond. Sodium chloride (NaCl) is nearly always quoted as an example of an ionic crystal although any combination of an alkali metal (with one electron outside a closed shell) and a halogen (with one electron missing from a closed shell) would do for an example. I shall remain faithful to sodium chloride. The chlorine atoms each gain an electron from a sodium atom and we are left with positive sodium ions Na^+ and negative chloride ions Cl^-. The ions attract one another – an electrostatic Coulomb attraction – and are therefore bound together in a solid lattice. There are two ways of defining the cohesive energy of an ionic crystal – the energy needed to separate the crystal into ions or the energy needed to separate it into the neutral atoms. For sodium chloride the cohesive energy (into ions) per ion pair is ~8 eV while the cohesive energy (into atoms) per ion pair is ~6 eV. More about ionic crystals will be found in chapter 3.

3. The covalent bond. A chlorine atom will take up an electron to give Cl^- if there are any spare electrons around or if it can steal one from a passing sodium atom, say. What happens when two chlorine atoms come into contact? The chlorine atoms are identical. There is no reason why one of the chlorines should be able to remove an electron forcibly from the other. But chlorine molecules, Cl_2, do exist. They make up chlorine gas. What happens – to put it very crudely – is that the chlorine atoms share their outer electrons which then form molecular orbitals around both atoms. A covalent bond has been formed. The charge density in this bond is high in the region between the atoms. The chlorine atom has only one missing electron and forms one covalent bond. Carbon has four electrons in its outer shell and this shell needs eight electrons to complete it. Carbon can form four covalent bonds and these bonds are directional. For carbon, in the form of diamond, these four bonds point to the corners of a tetrahedron. Silicon and germanium both form covalent bonds and crystallize in the diamond structure but here the covalent bonds are weaker than those in diamond. Diamond is the hardest material known. The cohesive energy for diamond is ~7 eV per atom and for silicon it is ~3.7 eV per atom.

4. The Van der Waals bond. A crystal in which the main bonding mechanism is the Van der Waals bond is solid argon. Argon, one of the noble gases, has atoms which are very unreactive as each one already has a closed-shell structure. The atoms in solid argon do not form ionic, covalent, or metallic bonds with their neighbours. Yet something holds the crystal lattice together. These bonding forces are not very strong – argon is only a solid at temperatures below 84 K. It was Van der Waals who first suggested a possible bonding mechanism. The electron charge-cloud around the positively charged nucleus can oscillate rapidly so that we have effectively a fluctuating dipole moment on the atom. A set of fluctuating dipoles will interact in such a way that the overall effect is one of attraction.[1] Many crystals of organic molecules are also bound together in this way. The internal bonds of the molecules are strong covalent bonds but the only forces between the molecules are Van der Waals forces.

TABLE 1.2

	nearest-neighbour spacing	melting point	structure	cohesive energy
Argon	3.8 Å	84 K	f.c.c.	0.08 eV/atom
Xenon	4.4 Å	161 K	f.c.c.	0.17 eV/atom

If all crystals were held together solely by one or other of these four types of bonds matters would be comparatively simple. However, *all* real bonds are a mixture of bonding types. I have given sodium chloride as an example of an *ionic* crystal but it will be obvious that Van der Waals forces operate in all crystals and so the bonding mechanism in sodium chloride cannot be purely ionic. Then there is a whole range of solids including the ceramics which are held together by bonds that are partly ionic and partly covalent. We shall now very sketchily consider the 'degree of ionicity' or 'per cent ionic character' of some of these bonds.

When an element readily forms anions – negative ions – we say that it is electronegative. It is possible to give each element a number called its electronegativity, which is a measure of this tendency to form negative ions. (Anyone who is enthusiastic enough to want to know how this number is arrived at should try reading a chemistry text such as Pauling.[2] Pauling suggested that the following formula should apply to a bond between an atom of type A and one of type B.

$$\text{Per cent ionic character} = 100\left[1 - \exp\left\{-\left(\frac{x_A - x_B}{2}\right)^2\right\}\right]$$

where x_A and x_B are the electronegativities of the two atoms.

Other people have since used different formulae and there is some experimental evidence. Unfortunately, experimental measurements on such things as the interaction of the nucleus with the electronic charge-cloud are not trivial to interpret. As an example of the numbers involved: The value of the per cent ionic character of zinc sulphide is usually quoted as around 65 per cent. The value for sodium chloride is given as between about 90 per cent and 100 per cent.

Crystal lattices

Before we are in a position to deal with real crystals we must first look at the Bravais space lattices. These are *mathematical* arrays of an infinite number of *points* in space such that the arrangements of points about a given point is exactly the same as the arrangement of points about any other point. From symmetry considerations Bravais showed that there were only fourteen distinct space lattices in three dimensions. You will be able to find pictures of all fourteen lattices in the standard texts. Four of these lattices – the only ones which will be mentioned explicitly in the rest of this book – are shown in Fig. 1.1. To describe the structure of any real crystal we talk in terms of these space lattices. Obviously it is easier to say a 'face-centred cubic lattice' than to write down the coordinates of the points or even to draw a picture. The four lattices shown in Fig. 1.1 will serve to describe many of the more common crystal structures.

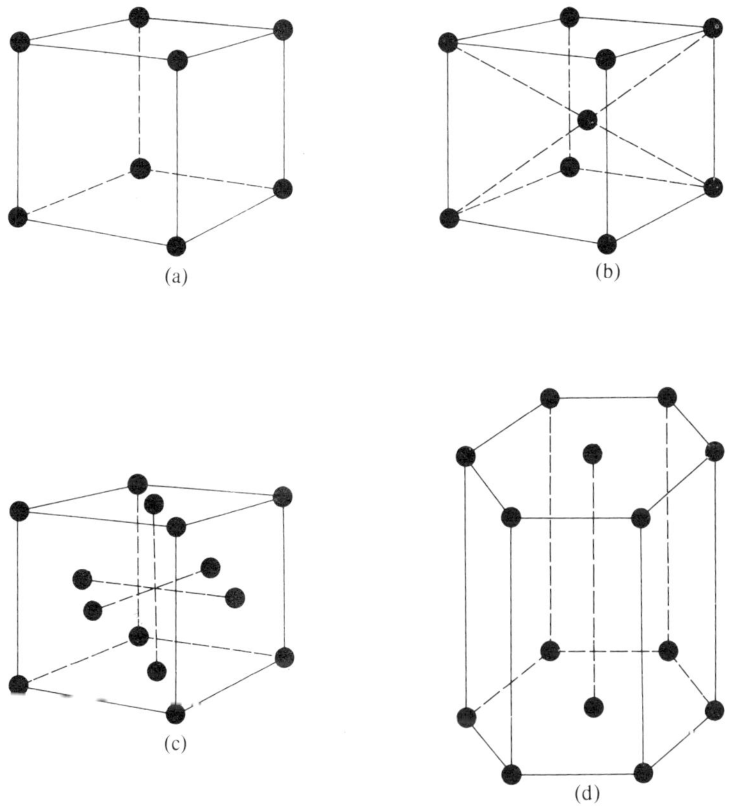

Fig. 1.1. A few of the (fourteen possible) Bravais space lattices. Shown are the conventional unit cells which, except for the simple cubic lattice, are not primitive unit cells. (a) Simple cubic; (b) body-centred cubic; (c) face-centred cubic; (d) hexagonal.

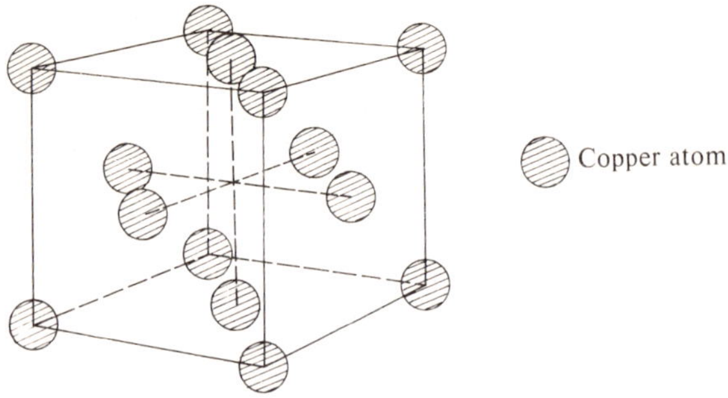

Fig. 1.2. The crystal structure of copper. The copper atom is drawn as a small sphere for clarity. It would perhaps be more realistic to have the spheres large enough to touch.

Now let us move on to the *crystal* lattices. It is here that many people begin to get confused because there is often more than one way of describing a crystal structure especially if the crystal contains more than one element. If I say that copper crystallizes in a face-centred cubic (f.c.c.) structure I am unlikely to confuse you. One simply draws a picture of the f.c.c. lattice from Fig. 1.1 and on each lattice point one puts a copper atom. This gives the picture in Fig. 1.2 and a perfect copper crystal is in fact like this except of course that it will be extended in all directions.

Figure 1.3 shows a small portion of a sodium chloride crystal. (You

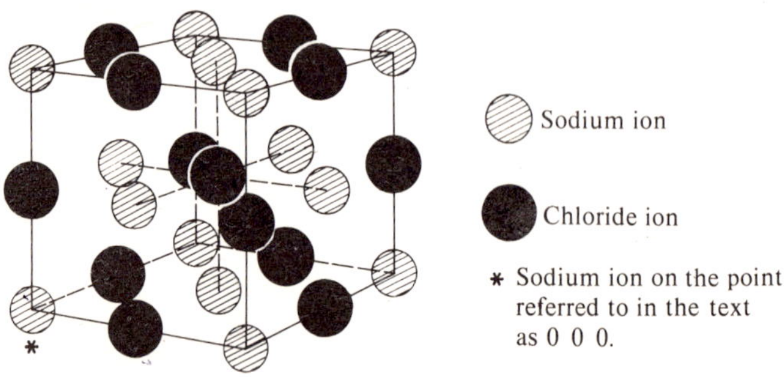

Fig. 1.3. The structure of sodium chloride. It would be more realistic to draw the ions touching and to make the sphere representing the chloride ion much larger than the one representing the sodium ion.

saw earlier that this particular crystal is ionic but in the text I shall use the word 'atom' as the precise nature of the bonding mechanism is not being considered here.) How can you describe this crystal lattice to someone in such a way that he can be left in no doubt of the correct structure?

1. You could say that the sodium atoms and the chlorine atoms are arranged on a simple cubic lattice occupying alternate sites so that each atom has, for nearest neighbours, atoms of the other element.

This may be an unambiguous description but would not find favour with crystallographers. They might tell you that the sodium chloride lattice does not have the translational symmetry properties of a simple cubic lattice (but see p. 14). What this amounts to is that it is not a good idea to say that a lattice site is occupied by *either* one type of atom *or* by another.

Look at the sodium atoms in the figure. You can see that they are on a f.c.c. lattice. Now look at the chlorine atom in the centre of the piece of the lattice shown. It is surrounded by six sodium atoms which can be said to be at the centre of an octahedron (Fig. 1.4). The chlorine atom is said to occupy an octahedral site. If you think of Fig. 1.3 as extended in all directions you will see that all the chlorine atoms are in equivalent positions.

2. The sodium chloride lattice can then be described as a f.c.c. lattice of sodium atoms which has all the octahedral sites occupied by chlorine atoms.

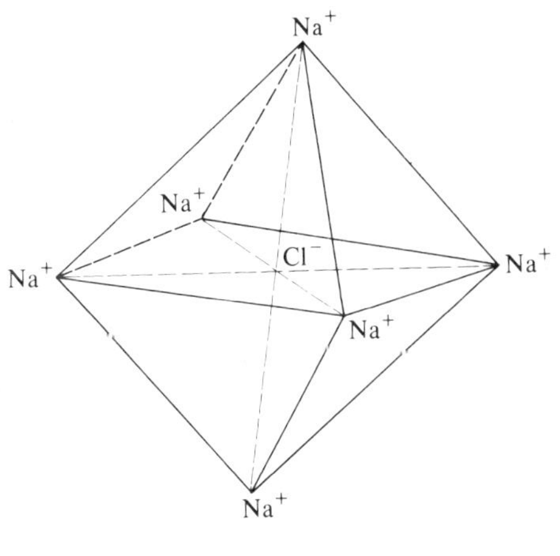

Fig. 1.4. The chloride ion in sodium chloride is at the centre of an octahedron which has sodium ions at the corners.

It can be seen from Fig. 1.3 that the sodium atoms and chlorine atoms form identical patterns. If all the sodium atoms were replaced by chlorine atoms and vice versa we should be left with the same crystal pattern.

3. Thus the sodium chloride lattice can equally well be described as a f.c.c. lattice of chlorine atoms with all the octahedral sites occupied by sodium atoms.

We now come to what is perhaps the most useful description of the sodium chloride lattice. This description involves a new concept – that of the basis set.

4. We can generate a sodium chloride crystal by replacing *each point* of the Bravais f.c.c. lattice by a pair of *atoms* – one sodium atom and one chlorine atom arranged in a special way. If we consider the sodium atom at the position labelled 0 0 0 on the diagram, then the corresponding chlorine atom is the one located at $\frac{1}{2}A$ $\frac{1}{2}A$ $\frac{1}{2}A$, where A is the length of the side of the cube shown in Fig. 1.3. Such a pair of atoms is a basis set for the sodium chloride crystal.

Generally, the *basis set is defined* as a group of atoms in a particular arrangement which must be placed on each point of a Bravais lattice in order to build up a crystal. Obviously a basis set must contain at least one 'molecule' of the substance. The simplest basis set for solid argon, which crystallizes in the f.c.c. structure, will be one argon atom. Some elements, for example carbon (in diamond), have a simplest basis set of two atoms. For some organic crystals the basis set may consist of hundreds or thousands of atoms.

Why take this particular pair of atoms as the basis set? Any of the eight positions $\pm\frac{1}{2}A$ $\pm\frac{1}{2}A$ $\pm\frac{1}{2}A$ which are occupied by the chlorine atoms would do perfectly well. There are also, of course, chlorine atoms at $\pm\frac{1}{2}A$ 0 0, 0 $\pm\frac{1}{2}A$ 0, and 0 0 $\pm\frac{1}{2}A$. Why can't you use these? Well, you can if you wish. It is easier to visualize the crystal structure if you stick to chlorine atoms near to 0 0 0 rather than take, for example, the chlorine atom at $22\frac{1}{2}A$ 0 0 but any pair of atoms could, in principle, be used to give an unambiguous description of the crystal lattice. The reason why I quoted $\frac{1}{2}A$ $\frac{1}{2}A$ $\frac{1}{2}A$ will become clear, I hope, when we come to look at the lattice vectors and the unit cell.

The lattice vectors and the unit cell. Perhaps it would be best to consider first a simple two-dimensional space lattice as shown in Fig. 1.5.

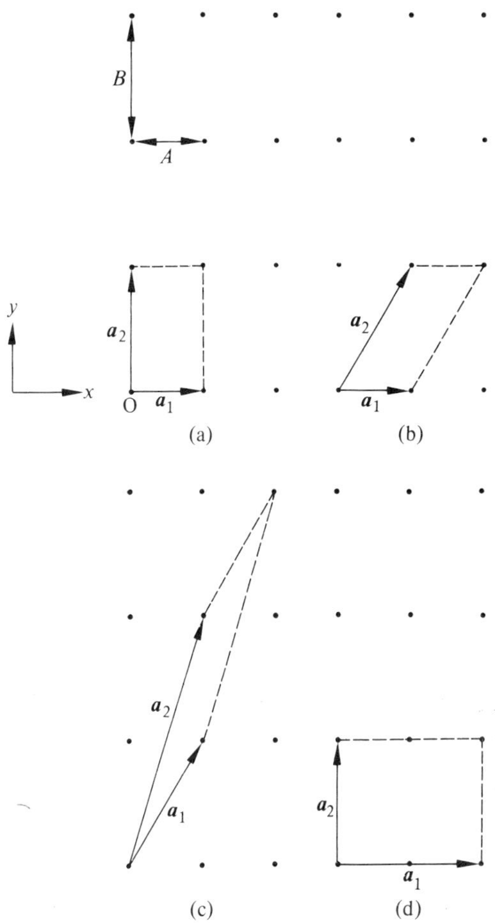

Fig. 1.5. A two-dimensional space lattice.

We can move from the point O to other points of this rectangular lattice by using a translation $\boldsymbol{L} = m\boldsymbol{a}_1 + n\boldsymbol{a}_2$ where m and n are integers. $\boldsymbol{a}_1$ and $\boldsymbol{a}_2$ are called the *basis vectors* and the general vector $\boldsymbol{L}$ is a *lattice vector*. The basis vectors $\boldsymbol{a}_1$ and $\boldsymbol{a}_2$ are said to be *primitive* if we can get from O to *every* other point of the lattice by using a vector $m\boldsymbol{a}_1 + n\boldsymbol{a}_2$.

The coordinates of the lattice points are $x = mA, y = nB$. The vector $\boldsymbol{a}_1$(a) points along the x-axis with magnitude A, and $\boldsymbol{a}_2$(a) points along the y-axis with magnitude B. Clearly (a) shows two primitive basis vectors. What about (b)? Well, as $a_1(b) = a_1(a)$ and $a_2(b) - a_1(b) = a_2(a)$ we can again cover all the grid with our lattice vectors. In (c) $2a_1(c) - a_2(c) = a_1(a)$ and $a_2(c) - a_1(c) = a_2(a)$. (b) and (c) show sets of primitive basis vectors. Now look at (d). No combination of $a_2(d)$ and $a_1(d)$ will give us $a_1(a)$. When we construct vectors $ma_1(a) + na_2(d)$ we shall never be able to reach the point that has coordinates A O or in fact any of those points which lie on the lines $x = (2N + 1)A$ where $N = 0,1,2$ etc. $a_1(d)$ and $a_2(d)$ are not primitive basis vectors.

Complete the parallelograms defined by the vectors a_1 and a_2. These parallelograms are the unit cells of the two-dimensional space lattice. Now suppose that the points of the mathematical lattice are replaced by circles so that we have a kind of two-dimensional monatomic crystal. Diagrams (a), (b), and (c) are such that inside each parallelogram we have a total of one circle ((a) has four quarter circles, etc.). Diagram (d) has the equivalent of two whole circles inside the rectangular unit cell – four times a quarter at the corners and two times one half along the sides. The 'crystal' lattice can be built up from these 'crystal' unit cells placed side by side. The three primitive unit cells, i.e. those formed from primitive basis vectors, each contain only one 'atom' whereas the non-primitive cell contains two. There are many more possible unit cells, both primitive and non-primitive (in fact an infinite number of each). All the non-primitive cells contain more than one 'atom' which means that the corresponding basis set is greater than that for the primitive cells.

The simplest primitive basis vectors for a f.c.c. Bravais lattice are shown in Fig. 1.6(a). These vectors define the positions of the nearest neighbour points relative to the point marked O for origin. There are eight nearest neighbours. The vectors a_1, a_2, and a_3 point to three of them. The vectors $-a_1$, $-a_2$, and $-a_3$ account for three more. The other two vectors can be expressed in terms of a_1, a_2, and a_3 (see above). The simplest primitive unit cell is the solid shown. All unit cells are parallelepipeds defined by the basis vectors. This particular unit cell is a rhomboid.

Now compare this mathematical lattice with the crystal lattice of Fig. 1.6(b). The crystal can be built up from unit cells (like those of the space lattice) and the matter they contain. Which atoms will the rhomboid contain? Atoms, of course, have a finite size – in fact they touch in the crystal – and one atom may overlap into more than one unit cell. Let us first say that we will consider an atom to be 'in' a cell if the point upon which it is positioned (the centre of the atom) is in the cell. Using this definition we see that the chlorine atom at

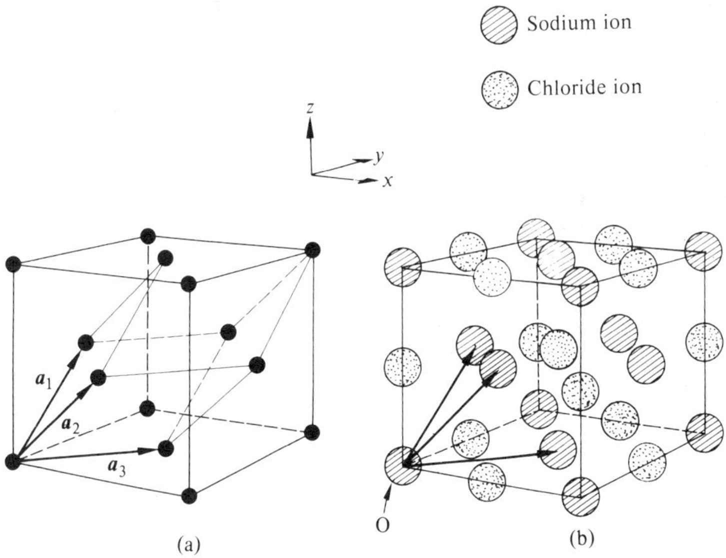

Fig. 1.6. (a) A primitive unit cell in a f.c.c. space lattice. (b) A primitive unit cell in the sodium chloride lattice.

$\frac{1}{2}A\ \frac{1}{2}A\ \frac{1}{2}A$ is certainly in the cell. But what about the sodium atom at 0 0 0? The way that I have drawn the cell means that there are eight sodium atoms associated with each cell and these are at the corners. Each sodium atom is shared among eight unit cells (although not evenly) and the net result is that there is the equivalent of one whole sodium atom 'in' the unit cell. Perhaps you don't like 'splitting the atom' and anyway we want to know which basis set of two whole atoms corresponds to a unit cell of the shape and orientation of our parallelepiped. Can a combination of the sodium atom at 0 0 0 and the chlorine atom at $\frac{1}{2}A\ \frac{1}{2}A\ \frac{1}{2}A$ be used? I have drawn the unit cell of the Bravais lattice so that its corners are on the lattice points and have assumed that the sodium atoms in the sodium chloride crystal are positioned at these lattice points. Neither assumption is necessary although for clarity the unit cells of Fig. 1.6(a) and (b) are the ones illustrated in books. Imagine our unit cell to be moved a *small* distance in the direction of a translational vector $(-1)\hat{x} + (-1)\hat{y} + (-1)\hat{z}$ where $\hat{x}, \hat{y}$, and $\hat{z}$ are unit vectors along the x-, y-, and z-axes. The sodium atom at 0 0 0 will now be totally included in the new unit cell and the other seven will be excluded. Perhaps it is now clear why I chose the basis set of atoms as the sodium atom at 0 0 0 and the chlorine atom at ½A ½A ½A.

You have seen that the simplest *primitive* unit cell of the f.c.c. lattice is *not* a cube. What about the set of points which form a cube and that keep appearing in the diagrams, for example in Fig. 1.6? This is one unit cell although it is not a primitive one. The atoms 'in' this unit cell can be used as a basis set on a now *simple cubic* lattice. The lattice is no longer the primitive lattice, i.e., the one built up from the primitive basis vectors. The atomic basis now consists of four 'molecules' of sodium chloride – the four sodium atoms at 0 0 0, ½A ½A 0, ½A 0 ½A, 0 ½A ½A and the four chlorine atoms at ½A 0 0, 0 ½A 0, 0 0 ½A, ½A ½A ½A. (Again I have moved the unit cell slightly so that no atoms are on the boundary.) We now have a fifth description of the sodium chloride lattice.

5. The sodium chloride lattice can be described as a simple cubic lattice with an atomic basis of eight atoms as shown in Fig. 1.7.

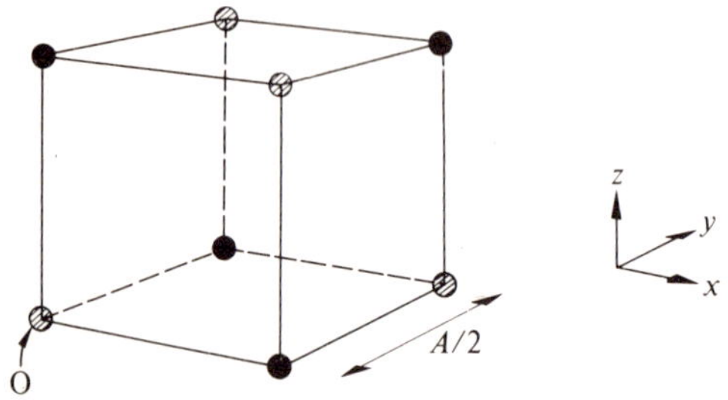

Fig. 1.7. A basis set of a sodium chloride lattice – to be arranged on a simple cubic lattice with side of the cube A.

The unit cell in the shape of a small cube with the corners coinciding with atomic positions is the *conventional* unit cell. The reasons for this are fairly obvious. It is so much easier for us all to imagine cubes arranged on a simple cubic lattice than to try to deal with rhomboids. The Cartesian axes are practically second nature to us. Also this particular choice of unit cell emphasizes the symmetry properties of the crystal.

(I have been giving the positions within the unit cell in terms of Cartesian coordinates, e.g., ½A ½A 0. You will find that the position within the unit cell is often given in dimensionless numbers as fractions of the side of the unit cell measured along the basis vectors which may not necessarily be at right angles to one another. The conventional unit cube of the f.c.c. lattice has basis vectors along Cartesian directions. Therefore we could write the position ½A ½A 0 as ½ ½ 0 *within the conventional unit cell*. However, *within* the primitive unit cell of Fig. 1.6 the coordinates would be given as 0 1 0 which means $0\boldsymbol{a}_1 + 1\boldsymbol{a}_2 + 0\boldsymbol{a}_3$.)

The subject of the sodium chloride structure is now practically closed. I must just mention that compounds which crystallize in this form are said to have the ROCKSALT STRUCTURE. Examples of such solids are compounds of the alkali metals (except sometimes caesium) with a halogen, e.g., lithium fluoride, potassium iodide; and compounds of the alkaline earth metals with oxygen, e.g., magnesium oxide, calcium oxide.

Figure 1.8 is a picture of a zinc sulphide crystal. In analogy with description (4) of the sodium chloride lattice I shall describe the structure of zinc sulphide as follows.

The zinc sulphide crystal is a f.c.c. lattice with a basis of two atoms such as the zinc atom at 0 0 0 and the sulphur atom at ¼ ¼ ¼. The positions are given in terms of the conventional unit cell which is the one shown in Fig. 1.8.

It can be seen from the diagram that each sulphur atom is surrounded by four nearest-neighbour zinc atoms which are at the corners of a tetrahedron. So we could say that the structure is that of a f.c.c. lattice of zinc atoms with half the tetrahedral sites occupied by sulphur atoms.

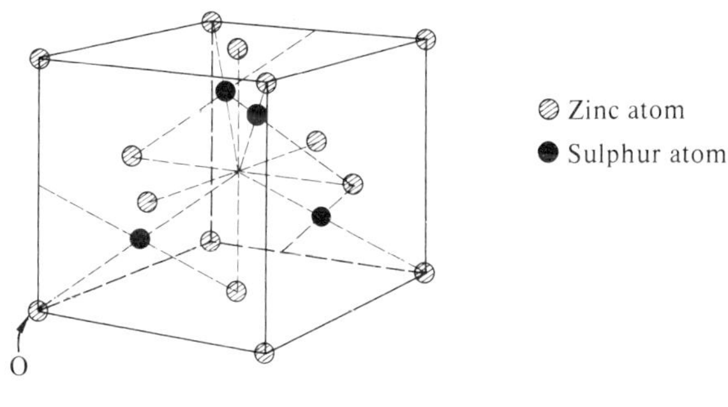

Fig. 1.8. Zinc sulphide.

Or we could say that it is a f.c.c. lattice of sulphur atoms with half the tetrahedral sites occupied by zinc atoms.

Substances which crystallize in a similar way to zinc sulphide are said to have the ZINC BLENDE STRUCTURE. Examples of such substances are cuprous chloride (copper chloride), beryllium oxide, zinc oxide and silicon carbide.

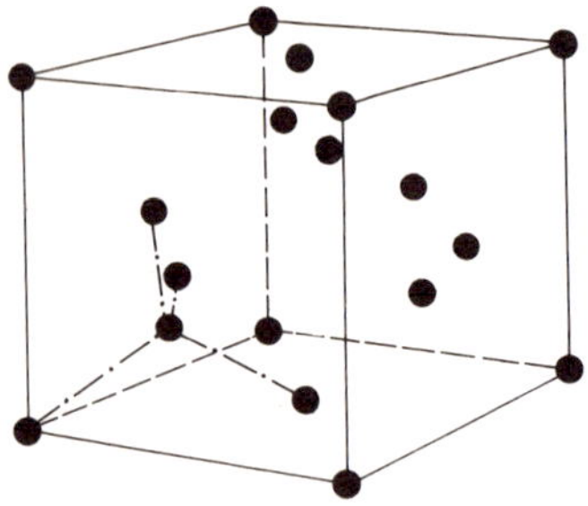

Fig. 1.9. The diamond structure.

The crystal structure of diamond is shown in Fig. 1.9. You will see that it is the same as the zinc blende structure except that the basis consists of two identical carbon atoms. Although I would hesitate to say that diamond crystals were common, the DIAMOND STRUCTURE occurs frequently in textbooks on solid-state physics. This is because the semiconductors silicon and germanium crystallize in this way. Each carbon atom in the crystal has four nearest neighbours arranged at the corners of a tetrahedron. Those of you who have dabbled to any extent in chemistry will be aware that the carbon atom likes to form covalent bonds in these directions. The molecule CH_4 (methane) is a carbon atom surrounded tetrahedrally by four hydrogen atoms. Carbon does crystallize in another form – graphite – about which I have nothing to say beyond the fact that you can find pictures of it elsewhere.

Unfortunately (for those people who like their geometry simple) no common element crystallizes in a simple cubic pattern. However, the caesium chloride lattice (see Fig. 1.10) has the translational symmetry properties of a simple cubic lattice. It can be described as a simple cubic lattice with a basis of two atoms – one at 0 0 0 and the other at ½ ½ ½. The CAESIUM CHLORIDE STRUCTURE is shared by caesium bromide and caesium iodide.

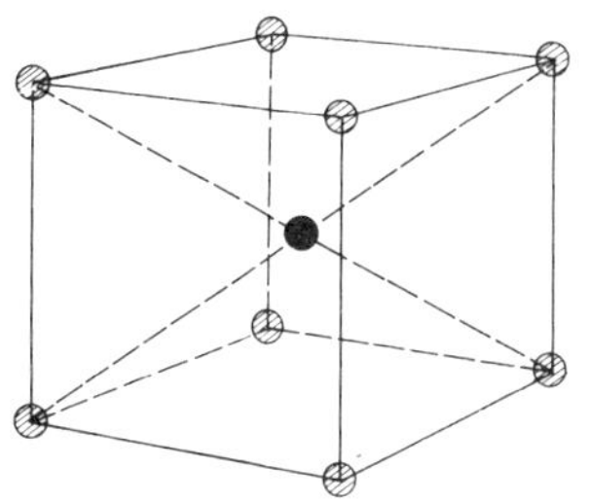

Fig. 1.10. Caesium chloride.

I shall mention one more crystal structure which is favoured by many metals – the HEXAGONAL CLOSE-PACKED STRUCTURE.

The conventional unit cell is shown in Fig. 1.11(a), while (b) shows the layers of atoms as seen from the top of the unit cell. (If the second layer were extended outside the unit cell you would see that it has exactly the same pattern as the other two layers but is displaced in its own plane.) The crosses representing the positions of the second layer are in the 'centres' of triangles of dots representing the first layer. The position of the 'centre' of a triangle is shown more clearly in (c). RC is two-thirds of RS. Similarly, QC is two-thirds of QU and PC is two-thirds of PT. The line VC drawn parallel to QP will divide RQ in the same ratio as it divides RS. Therefore

$$RV = \tfrac{2}{3} RQ.$$

What is VC in terms of QP? By similar triangles, $VC/QS = RV/RQ = \tfrac{2}{3}$, and as QS is QP/2 we have

$$VC/QP = \tfrac{1}{3}.$$

So, using the basis vectors a_1, a_2, and a_3 shown in Fig. 1.11(a), we have the coordinates of the atom labelled X as $\tfrac{2}{3}\ \tfrac{1}{3}\ \tfrac{1}{2}$. The $\tfrac{1}{2}$ is there because the second layer is half-way between the first and third layers of atoms.

Therefore, the hexagonal close-packed structure has the translational symmetry of a hexagonal lattice with a basis of two atoms, one at 0 0 0 and the other at $\tfrac{2}{3}\ \tfrac{1}{3}\ \tfrac{1}{2}$. The unit cell is not close-packed unless there is a special ratio between the lengths A and C. This brings us nicely to the next topic.

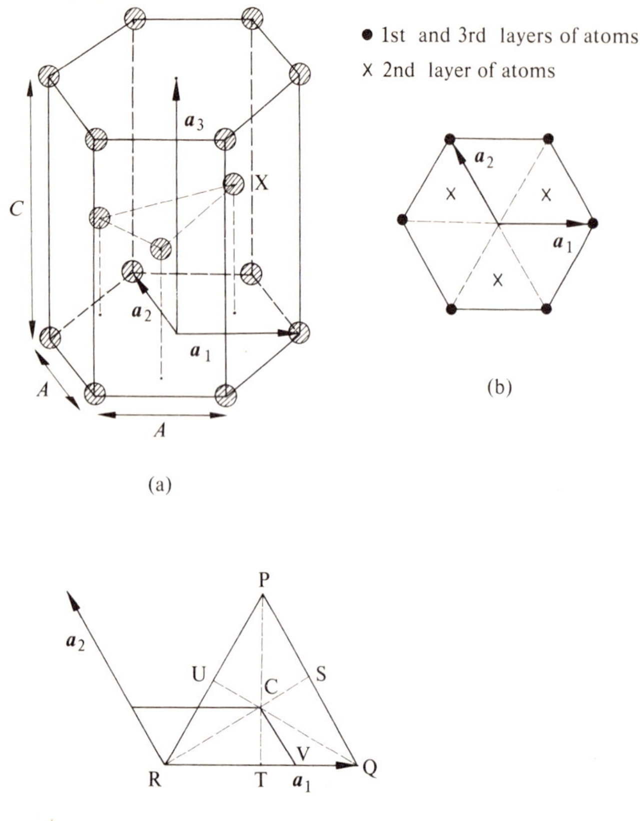

Fig. 1.11. (a) Conventional unit cell for a hexagonal close-packed structure. (If the three atoms in the centre are removed we are left with the usual hexagonal unit cell – see Fig. 1.1(d).) (b) The unit cell viewed from above. (c) The position of the 'centre' of a triangle.

Packing in crystals

The actual crystal structure that any substance will take depends on many factors. Factors influencing the structure of ionic crystals will be mentioned in chapter 3. When the bonding is mainly covalent the 'bond directions' – positions of the nearest neighbours – will be dependent on the atomic structure of the constituents. However, one of the main considerations which dictate the form of metallic crystals and the noble-gas solids is the packing ratio of the different possible lattice types. The nature of the metallic bond and the Van der Waals bond is such that binding will be strongest if each atom in the solid has as many neighbours as close as possible (within the restriction on nearest-neighbour spacing). (The symmetry properties of molecules which are held together by Van der Waals forces will influence the structure. Such complications do not arise for the elemental noble-gas solids. The relative sizes of the different metal atoms will have an effect on the crystal structure of alloys.) How do we set about calculating which lattice types are packed tightest? We could simply work out the number of atoms per unit volume in different crystal lattices and we would finish up with a value which depended on the nearest-neighbour spacing and on our definition of unit volume.

Instead we work out a dimensionless number called the filling factor.

$$\text{Filling factor} = \frac{\text{volume of space occupied by the atoms (in the unit cell)}}{\text{total volume (of unit cell)}}.$$

We treat the atoms as hard spheres. As we are dealing with elemental solids all the atoms will be identical. Let us call the atomic radius R.

1. Consider first a simple *hexagonal* point lattice as shown in Fig. 1.1(d). Replace the points by hard spheres of radius R. Obviously the spheres must touch or the lattice would collapse. The bottom layer of atoms will look like Fig. 1.12(a).

$$A = 2R.$$

The two layers will touch so $C = 2R$.

How many atoms are in the unit cell? The twelve atoms at the corners are all equally shared between six unit cells. They contribute a total of two atoms. The atoms in the centres of the top and bottom of the cell are each shared between two unit cells. They contribute a total of one atom to the cell. Therefore, in the conventional unit cell we have three atoms.

$$\begin{aligned}\text{The volume of the cell} &= C \times \text{area of hexagon}\\ &= C \times 6 \times \tfrac{1}{2}A \cdot A \sin 60^\circ\\ &= CA^2 \frac{\sqrt{3}}{2} 3.\end{aligned}$$

Substituting for A and C in terms of R we get

$$\text{Volume of unit cell} = 12\sqrt{(3)}R^3.$$

$$\text{Filling factor} = \frac{3 \times \tfrac{4}{3}\pi R^3}{12\sqrt{(3)}R^3} = \frac{\pi}{3\sqrt{3}} \text{ or 60 per cent.}$$

2. The *hexagonal close-packed* lattice has a conventional unit cell as shown in Fig. 1.11(a). Replace each point by a hard sphere of radius R. In order that this lattice should be truly close-packed we need a special relationship between C and A. $A = 2R$ as before but now C is limited by the fact that the distance between the two basis atoms – one at 0 0 0 and the other at $\frac{2}{3}\ \frac{1}{3}\ \frac{1}{2}$ – must be equal to $2R$. Refer to Fig. 1.12(b). This distance is marked as D.

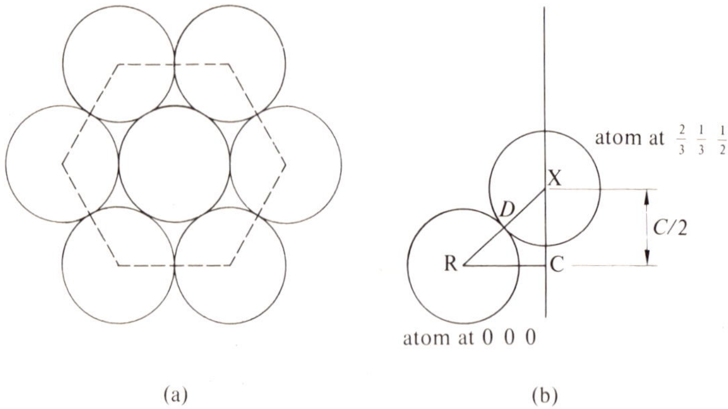

Fig. 1.12. Packing in the hexagonal close-packed lattice.

$$D^2 = (\mathrm{RC})^2 + \left(\frac{C}{2}\right)^2.$$

From Fig. 1.11(c)

$$(\mathrm{RC})^2 = (\tfrac{2}{3}\mathrm{RS})^2 = (\tfrac{2}{3}A \sin 60°)^2 = \tfrac{1}{3}A^2.$$

Therefore

$$D^2 = \frac{A^2}{3} + \frac{C^2}{4}.$$

As $D = 2R$ and $A = 2R$ we can write $D = A$ and we get

$$A^2 = \frac{A^2}{3} + \frac{C^2}{4} \quad \text{or} \quad C = \tfrac{8}{3}A^2, \quad \frac{C}{A} = 1.633.$$

For most real hexagonal crystals C/A is in fact greater than 1.633 but these crystals are referred to as hexagonal close-packed (h.c.p.) provided that the ratio is not more than about ten per cent greater than the ideal of 1.633. Real lattices don't collapse if C/A is greater than this – atoms are not really hard spheres!

For an h.c.p. lattice, the number of atoms in the conventional unit cell is the three atoms of the hexagonal lattice plus the three at the centre, i.e., six in all. The volume of the cell is $CA^2\sqrt{3}/4$. Here $C = A\sqrt{(8/3)}$ and $A = 2R$. Therefore the volume is $R^3\,24\sqrt{2}$.

$$\text{Filling factor} = \frac{6 \times \tfrac{4}{3}\pi R^3}{24\sqrt{(2)}R^3} = \frac{\pi}{3\sqrt{2}} \text{ or 74 per cent.}$$

3. The *face-centred cubic* lattice. Here the limiting factor is that the face diagonal of the conventional unit cell must be equal to $4R$ (see Fig. 1.13(a)). We have

$$\sqrt{(2)}A = 4R.$$

The atoms on the faces of the cube contribute half their volume to the unit cell. The atoms at the corners contribute one-eighth of their volume. Therefore, the volume of atoms within the unit cell is

$$(6 \times \tfrac{1}{2} + 8 \times \tfrac{1}{8})\,\tfrac{4}{3}\pi R^3 = \tfrac{16}{3}\pi R^3.$$

The volume of the cell is A^3 or $16\sqrt{(2)}R^3$.

$$\text{Filling factor} = \frac{\tfrac{16}{3}\pi R^3}{16\sqrt{(2)}R^3} = \frac{\pi}{3\sqrt{2}}$$

which is exactly the same as for the h.c.p. structure.

4. The *body-centred cubic* lattice. Here the limiting factor is that the body diagonal of the cube must be equal to $4R$ (see Fig. 1.13(b)).

$$\sqrt{(3)}A = 4R.$$

The volume of atoms within the unit cell is

$$(1 \times 1 + 8 \times \tfrac{1}{8})\,\tfrac{4}{3}\pi R^3 = \tfrac{8}{3}\pi R^3.$$

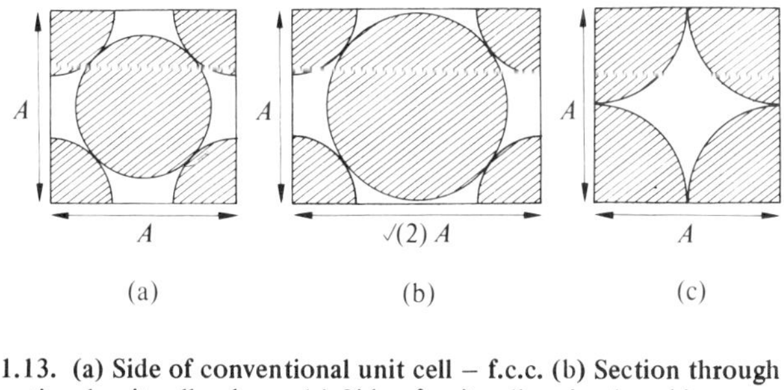

Fig. 1.13. (a) Side of conventional unit cell – f.c.c. (b) Section through conventional unit cell – b.c.c. (c) Side of unit cell – simple cubic.

The volume of the cell is A^3 or $\frac{64}{3\sqrt{3}} R^3$.

$$\text{Filling factor} = \frac{\frac{8}{3}\pi R^3}{64R^3/(3\sqrt{3})} = \frac{\sqrt{3}}{8}\pi = 68 \text{ per cent.}$$

5. The *simple cubic* lattice. The side of the cube, A, must be equal to $2R$ (see Fig. 1.13(c)).

$$\text{Filling factor} = \frac{(8 \times \frac{1}{8})\ \frac{4}{3}\pi R^3}{8R^3} = \frac{\pi}{6} \text{ or 52 per cent.}$$

We can see from examples (1) to (5) that the h.c.p. and f.c.c. lattices are the most closely packed ones. Many pure metals crystallize in one or other of these forms.

In the diamond lattice packing considerations are not very important. It is left as a problem for the reader to calculate the filling factor for an array of hard spheres on a diamond structure lattice.

Reciprocal lattice vectors

We now come to a topic which, at first sight, appears to be a purely mathematical abstraction. One way in which reciprocal lattice vectors help in the study of solids will become evident in the next sections and further uses will appear in later chapters. But for the moment I must ask you to bear with me.

The basis set of reciprocal lattice vectors are defined by the equation

$$\boldsymbol{a}_i \cdot \boldsymbol{b}_j = 2\pi\delta_{ij} \qquad (\delta_{ij} = 0 \text{ if } i \neq j, \delta_{ij} = 1 \text{ if } i = j)$$

where the $\boldsymbol{a}$'s are the basis vectors of the direct lattice. The $\boldsymbol{b}$'s are then called the *fundamental vectors of the reciprocal lattice*. The $\boldsymbol{b}$'s will vary for any given crystal structure depending on our choice of $\boldsymbol{a}$'s. We saw before that (in two or more dimensions) we could choose any number of different sets of basis vectors – some primitive and others not. We usually find $\boldsymbol{b}$'s corresponding to the $\boldsymbol{a}$'s of either the simplest primitive unit cell or the conventional unit cell. For example, the basis vectors corresponding to the conventional unit cell of a f.c.c. lattice are those corresponding to a primitive unit cell of a simple cubic lattice. Thus the $\boldsymbol{a}$'s (conventional cell) and hence the $\boldsymbol{b}$'s (conventional cell) are at right angles to one another; the $\boldsymbol{a}$'s (primitive) and $\boldsymbol{b}$'s (primitive) are not.

There is no obvious reason as yet for the factor of 2π in the equation. In fact many authors (especially crystallographers) prefer to define the reciprocal lattice vectors without this factor. You will see later that we shall use reciprocal lattice vectors in such a way that if the factor 2π is not included in the definition we shall find ourselves always writing $2\pi\boldsymbol{b}$. It is simpler, therefore, to put the 2π into the definition and have done with it. It is important to realize, though, that many people do prefer the other way so if you attempt problems in books you may find that you have an answer 'wrong' by 2π, $4\pi^2$, or $8\pi^3$.

What form do these $\boldsymbol{b}$-vectors take? First of all consider a one-dimensional lattice. Here we have

$$\boldsymbol{a}_1 \cdot \boldsymbol{b}_1 = 2\pi$$

or $\quad a_1 b_1 = 2\pi \qquad b_1 = 2\pi/a_1.$

The dimensions are clearly those of length^{-1}.

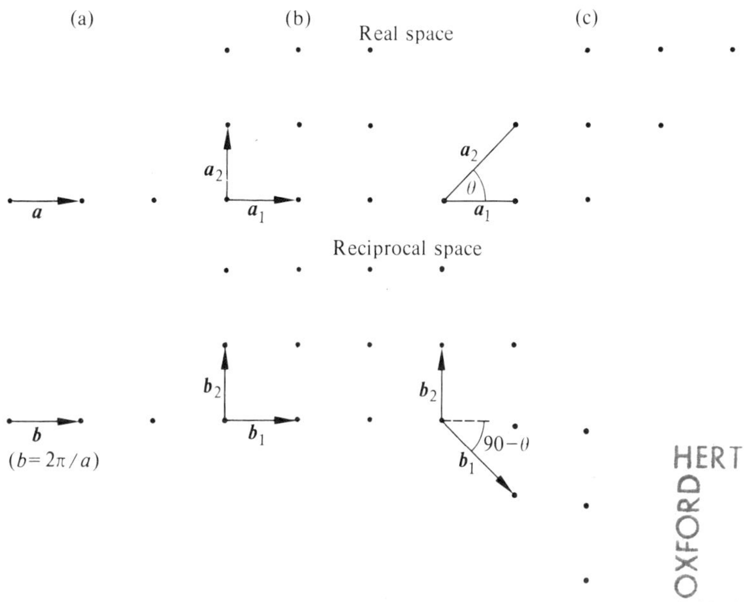

Fig. 1.14. Reciprocal lattices.

In two dimensions we have

$$\boldsymbol{a}_1 \cdot \boldsymbol{b}_1 = 2\pi \qquad \boldsymbol{a}_2 \cdot \boldsymbol{b}_2 = 2\pi$$

$$\boldsymbol{a}_1 \cdot \boldsymbol{b}_2 = 0 \qquad \boldsymbol{a}_2 \cdot \boldsymbol{b}_1 = 0.$$

From these equations we see that $\boldsymbol{b}_1$ must be perpendicular to $\boldsymbol{a}_2$ and that $\boldsymbol{b}_2$ must be perpendicular to $\boldsymbol{a}_1$. See Fig. 1.14(b) for a picture of a two-dimensional rectangular lattice. If $\boldsymbol{a}_1$ and $\boldsymbol{a}_2$ have an angle $\theta = 90°$

between them, the conditions are automatically fulfilled if $\boldsymbol{b}_1$ is in the same direction as $\boldsymbol{a}_1$ and $\boldsymbol{b}_2$ is in the same direction as $\boldsymbol{a}_2$. The reciprocal lattice – a set of points in reciprocal space which are connected to a given point by the vectors

$$G = n_1 \boldsymbol{b}_1 + n_2 \boldsymbol{b}_2$$

where n_1 and n_2 are integers – is also rectangular. The magnitudes of the vectors are given by $b_1 = 2\pi/a_1$, $b_2 = 2\pi/a_2$.

Now look at Fig. 1.14(c) which shows a general two-dimensional lattice. If the angle between $\boldsymbol{a}_1$ and $\boldsymbol{a}_2$ is θ then the angle between $\boldsymbol{b}_1$ and $\boldsymbol{b}_2$ is $180° - \theta$. The magnitudes of the reciprocal lattice vectors are

$$b_1 = \frac{2\pi}{a_1} \cos(90° - \theta), \quad b_2 = \frac{2\pi}{a_2} \cos(90° - \theta).$$

In three dimensions we can define the fundamental reciprocal lattice vectors as

$$\boldsymbol{b}_1 = \frac{2\pi \boldsymbol{a}_2 \times \boldsymbol{a}_3}{\boldsymbol{a}_1 \cdot (\boldsymbol{a}_2 \times \boldsymbol{a}_3)} \quad \text{etc.}$$

The term in the denominator is a scalar quantity and is equal to the volume of the unit cell. It is there for normalization purposes. Taking the scalar product of $\boldsymbol{a}_1$ with both sides of the equation we get

$$\boldsymbol{a}_1 \cdot \boldsymbol{b}_1 = \frac{2\pi \boldsymbol{a}_1 \cdot (\boldsymbol{a}_2 \times \boldsymbol{a}_3)}{\boldsymbol{a}_1 \cdot (\boldsymbol{a}_2 \times \boldsymbol{a}_3)} = 2\pi.$$

The vector product $\boldsymbol{a}_2 \times \boldsymbol{a}_3$ is perpendicular to both $\boldsymbol{a}_2$ and $\boldsymbol{a}_3$ and so

$$\boldsymbol{a}_2 \cdot \boldsymbol{b}_1 = \frac{2\pi \boldsymbol{a}_2 \cdot (\boldsymbol{a}_2 \times \boldsymbol{a}_3)}{\boldsymbol{a}_1 \cdot (\boldsymbol{a}_2 \times \boldsymbol{a}_3)} = 0$$

$$\boldsymbol{a}_3 \cdot \boldsymbol{b}_1 = \frac{2\pi \boldsymbol{a}_3 \cdot (\boldsymbol{a}_2 \times \boldsymbol{a}_3)}{\boldsymbol{a}_1 \cdot (\boldsymbol{a}_2 \times \boldsymbol{a}_3)} = 0$$

because the scalar product of two vectors at right angles depends on cos 90° which is zero. In three dimensions the reciprocal lattice is made up from points connected to some given point defined as the origin by the general vectors

$$G = n_1 \boldsymbol{b}_1 + n_2 \boldsymbol{b}_2 + n_3 \boldsymbol{b}_3.$$

By analogy with the two-dimensional case, we can see that the fundamental vectors for a three-dimensional simple cubic lattice are given by

$$b_1 = \left(\frac{2\pi}{a_1}\right)\left(\frac{a_1}{a_1}\right) \text{ etc.}$$

That is, b_1 has magnitude $= 2\pi/a_1$ and is in the same direction as a_1.

But what are the fundamental reciprocal lattice vectors corresponding to the primitive basis vectors of a f.c.c. lattice? If you turn back to Fig. 1.6(a) you will see that the vectors a can be described in Cartesian coordinates as

$$a_1 = (0)\hat{x} + (\tfrac{1}{2}A)\hat{y} + (\tfrac{1}{2}A)\hat{z}$$

$$a_2 = (\tfrac{1}{2}A)\hat{x} + (0)\hat{y} + (\tfrac{1}{2}A)\hat{z}$$

$$a_3 = (\tfrac{1}{2}A)\hat{x} + (\tfrac{1}{2}A)\hat{y} + (0)\hat{z}.$$

Let us first consider the term $a_1 \cdot (a_2 \times a_3)$. This is the volume of the primitive unit cell and is equal to both $a_3 \cdot (a_1 \times a_2)$ and $a_2 \cdot (a_3 \times a_1)$. We already know, from our studies of the different basis sets of the sodium chloride lattice, that the volume of the primitive unit cell is one-quarter of the volume of the conventional unit cell, i.e., $A^2/4$. We can check this from vector algebra.

Two vectors, v and w, have a scalar product

$$v \cdot w = v_x w_x + v_y w_y + v_z w_z.$$

Their vector product is

$$v \times w = (v_y w_z - v_z w_y)\hat{x} + (v_z w_x - v_x w_z)\hat{y} + (v_x w_y - v_y w_x)\hat{z}.$$

If $v = a_2$ and $w = a_3$,

$$\begin{aligned} a_2 \times a_3 &= (0 - \tfrac{1}{4}A^2)\hat{x} + (\tfrac{1}{4}A^2 - 0)\hat{y} + (\tfrac{1}{4}A^2 - 0)\hat{z} \\ &= (-\tfrac{1}{4}A^2)\hat{x} + (\tfrac{1}{4}A^2)\hat{y} + (\tfrac{1}{4}A^2)\hat{z} \end{aligned}$$

If now we write $v = a_1$ and $w = a_2 \times a_3$, we find

$$a_1 \cdot (a_2 \times a_3) = 0 + \frac{A^3}{8} + \frac{A^3}{8} = \frac{A^3}{4}$$

Then

$$b_1 = \frac{2\pi}{A^3/4}\{(-\tfrac{1}{4}A^2)\hat{x} + (\tfrac{1}{4}A^2)\hat{y} + (\tfrac{1}{4}A^2)\hat{z}\} = \frac{2\pi}{A}\{-\hat{x} + \hat{y} + \hat{z}\}$$

$$b_2 = 2\pi\frac{a_3 \times a_1}{a_2 \cdot (a_3 \times a_1)} \quad \text{which is} \quad \frac{2\pi}{A}\{\hat{x} - \hat{y} + \hat{z}\} \tag{1.1}$$

$$b_3 = 2\pi\frac{a_1 \times a_2}{a_3 \cdot (a_1 \times a_2)} \quad \text{which is} \quad \frac{2\pi}{A}\{\hat{x} + \hat{y} - \hat{z}\}.$$

To what sort of a lattice in reciprocal space do these fundamental vectors correspond? They represent a b.c.c. lattice as can be seen from Fig. 1.15. From the point at the centre of the 'conventional unit cell', the points at the corners can be defined by reciprocal lattice vectors

$$G = \boldsymbol{b}_1, \quad G = \boldsymbol{b}_2, \quad G = \boldsymbol{b}_3, \quad G = \boldsymbol{b}_4$$

$$G = -\boldsymbol{b}_1, \quad G = -\boldsymbol{b}_2, \quad G = -\boldsymbol{b}_3, \quad G = -\boldsymbol{b}_4.$$

But $\boldsymbol{b}_4$ is the vector

$$\frac{2\pi}{A}(\hat{x} + \hat{y} + \hat{z}) \quad \text{and this is equal to} \quad \boldsymbol{b}_1 + \boldsymbol{b}_2 + \boldsymbol{b}_3$$

which can be seen from the diagram or from eq. (1.1). Each point in the body-centred cubic (b.c.c.) lattice is surrounded by points in the same way as is the point labelled O. Therefore, if we can get to all the nearest neighbours by use of the translational vectors $\boldsymbol{b}_1, \boldsymbol{b}_2, \boldsymbol{b}_3$, we can reach every point in the lattice by a translation of the form

$$G = n_1\boldsymbol{b}_1 + n_2\boldsymbol{b}_2 + n_3\boldsymbol{b}_3.$$

The vectors $\boldsymbol{b}_1$, $\boldsymbol{b}_2$, and $\boldsymbol{b}_3$ are primitive vectors of a b.c.c. lattice in reciprocal space.

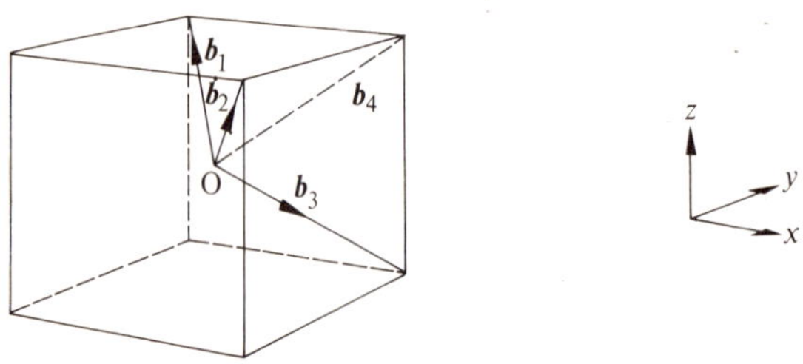

Fig. 1.15. A b.c.c. lattice in reciprocal space.

It has been shown that the fundamental vectors of the reciprocal lattice which correspond to the primitive basis vectors of a f.c.c. lattice in real space will generate a b.c.c. lattice in reciprocal space. We used the simplest primitive basis vectors of the f.c.c. lattice but the use of any primitive basis would have given the same result.

It can be shown that a b.c.c. lattice in real space will transform to a f.c.c. lattice in reciprocal space but the proof is left as an exercise for the reader.

So, for a given crystal and a given choice of unit cell, we can find the fundamental set of reciprocal lattice vectors and set up a reciprocal lattice. We can get from a given point to any other point by a translation of the form

$$\boldsymbol{G} = n_1 \boldsymbol{b}_1 + n_2 \boldsymbol{b}_2 + n_3 \boldsymbol{b}_3 \quad (n_1, n_2, n_3 \text{ integers}).$$

Now the vectors to the nearest neighbour points may not be those given by $(n_1 = \pm 1, n_2 = 0, n_3 = 0)$, $(n_1 = 0, n_2 = \pm 1, n_3 = 0)$, $(n_1 = 0, n_2 = 0, n_3 = \pm 1)$. The values of the integers will depend on the choice of unit cell, i.e., on our choice of the basis vectors $\boldsymbol{a}$. However, regardless of which values of n_1, n_2, and n_3 we may have to use to obtain them, the vectors to the nearest neighbours will be the same (in both magnitude and direction) for all sets of $\boldsymbol{b}$ derived from different primitive basis vectors $\boldsymbol{a}$. The *lattice* of points in reciprocal space *is the same* although it can be set up from different fundamental vectors.

This lattice will only change if our choice of unit cell is non-primitive. For instance, the conventional unit cell of the f.c.c. lattice is defined by vectors which are, in fact, primitive vectors of a simple cubic lattice of spacing A (see Fig. 1.1(c)). There are other possible unit cells of the f.c.c. lattice which are defined by other sets of primitive vectors of this same simple cubic structure. The defining vectors $\boldsymbol{a}$ of all these unit cells will transform to different sets of $\boldsymbol{b}$ which represent the same reciprocal lattice (in this case a simple cubic one).

We will study this in more detail. Consider Fig. 1.16 which shows a simple two-dimensional square lattice of spacing A and the reciprocal lattice – also square but with spacing $2\pi/A$. $\boldsymbol{a}_1\boldsymbol{a}_2, \boldsymbol{a}'_1\boldsymbol{a}'_2$ are sets of primitive basis vectors while $\boldsymbol{b}_1\boldsymbol{b}_2, \boldsymbol{b}'_1\boldsymbol{b}'_2$ are the corresponding fundamental reciprocal lattice vectors. In (a) I have used the shortest vectors $\boldsymbol{a}_1$ and $\boldsymbol{a}_2$, i.e., those that define positions of nearest-neighbour points to a reference point O. The corresponding vectors $\boldsymbol{b}_1$ and $\boldsymbol{b}_2$ are the shortest possible reciprocal lattice vectors.

$$\boldsymbol{G} = n_1 \boldsymbol{b}_1 + n_2 \boldsymbol{b}_2$$

will define a nearest-neighbour point (give $|\boldsymbol{G}|$ a minimum) where

$$n_1 = \pm 1, \quad n_2 = 0 \quad \text{or} \quad n_1 = 0, \quad n_2 = \pm 1.$$

In (b) I have used another set of primitive basis vectors. Now $\boldsymbol{b}'_2$ is perpendicular to $\boldsymbol{a}'_2$.

$$b'_1 = 2\pi/\{A \cos 45^\circ\} = 2\sqrt{2}\pi/A.$$

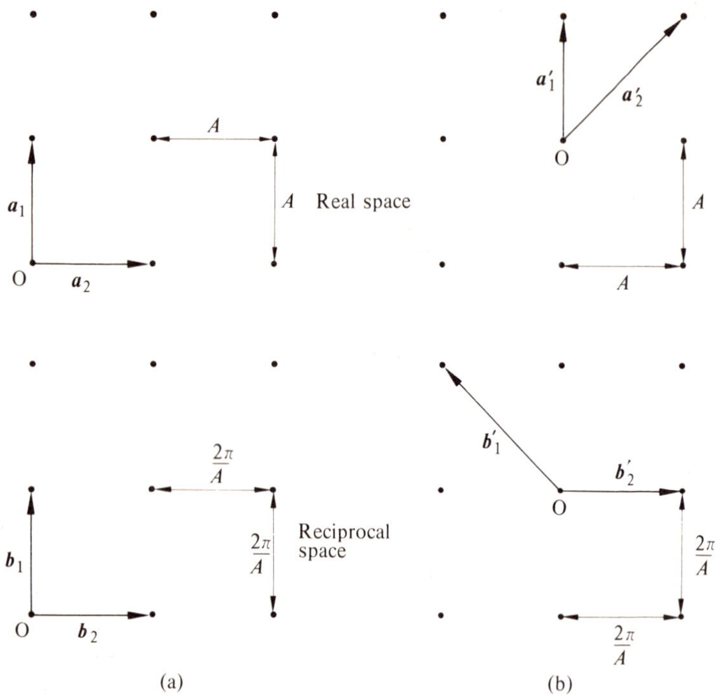

Fig. 1.16. This figure shows that different sets of primitive basis vectors will give the same reciprocal lattice.

Also $\boldsymbol{b}'_2$ is perpendicular to $\boldsymbol{a}'_1$.

$$b'_2 = 2\pi/\{(A/\cos 45^\circ)\cos 45^\circ\} = 2\pi/A.$$

$$\boldsymbol{G} = n_1\boldsymbol{b}'_1 + n_2\boldsymbol{b}'_2$$

will define a nearest-neighbour point when

$$n'_1 = 0, \quad n'_2 = \pm 1 \quad \text{or} \quad n'_1 = \pm 1 = n'_2.$$

The reciprocal lattice and the possible values of $\boldsymbol{G}$ are the same in both cases although to express any given $\boldsymbol{G}$ in terms of $\boldsymbol{b}_1$ and $\boldsymbol{b}_2$ will need different integers than the expression in terms of $\boldsymbol{b}'_1$ and $\boldsymbol{b}'_2$.

The Brillouin zone

In three dimensions we have a lattice represented by a set of points defined by

$$\boldsymbol{G} = n_1\boldsymbol{b}_1 + n_2\boldsymbol{b}_2 + n_3\boldsymbol{b}_3.$$

Suppose that we 'draw' the planes which are the perpendicular bisectors of all the reciprocal lattice vectors. Around the reference point O there will be a small 'volume' of reciprocal space which is bounded by planes that are the perpendicular bisectors of the vectors connecting O to near neighbours. This 'volume' is defined as the first Brillouin zone. For a simple cubic real lattice the reciprocal lattice is also simple cubic and the first Brillouin zone is a cube.

For a f.c.c. real lattice the reciprocal lattice is b.c.c. Each point in a b.c.c. lattice has eight nearest neighbours. The Brillouin zone is approximately a 'solid' with eight faces but it is modified because the corners are cut by the planes that are the bisectors of the six next-nearest neighbours.

A real b.c.c. lattice has a reciprocal lattice which is f.c.c. Each point in an f.c.c. lattice has twelve nearest neighbours. Here the first Brillouin zone is a 'solid' with twelve faces – a regular dodecahedron.

It is possible to define a second Brillouin zone as the volume, or rather bits of volume, lying just outside the first zone and bounded by another set of planes. Also one can define a third, a fourth, and so on, but the first zone is the one that we shall come across most frequently. All zones have the same volume in reciprocal space.

Figure 1.17 shows the first three Brillouin zones for a square

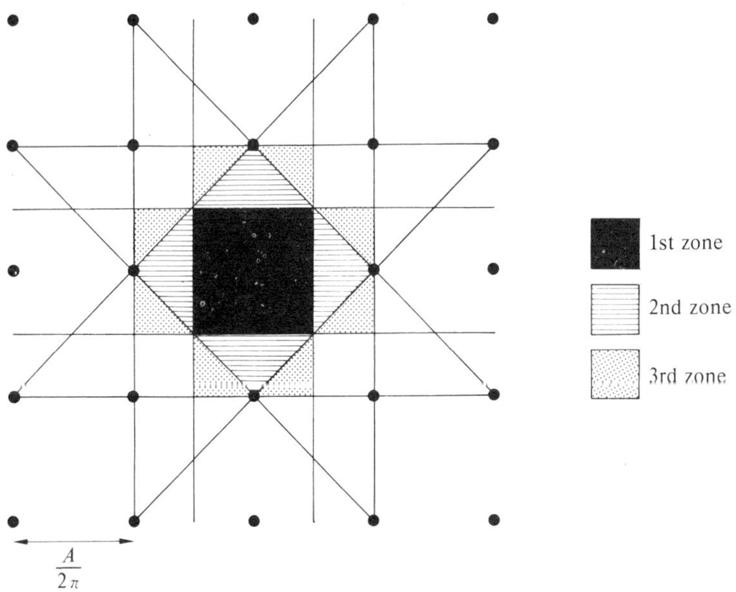

Fig. 1.17. A reciprocal lattice corresponding to a real square lattice of side A.

reciprocal lattice. The planes bisecting the vectors

$$\pm\frac{2\pi}{A}\hat{x}, \quad \pm\frac{2\pi}{A}\hat{y}, \quad \pm\frac{2\pi}{A}\hat{x}\pm\frac{2\pi}{A}\hat{y}, \quad \pm\frac{4\pi}{A}\hat{x}, \quad \text{and} \quad \pm\frac{4\pi}{A}\hat{y}$$

are shown. These planes are sufficient to define the first three zones. Other planes will be too far from the origin to cut through the zones shown. Each zone has the same area in reciprocal space ($4\pi^2/A^2$ here).

The periodic structure of the crystal lattice is responsible for the fact that all distinct solutions of a wave equation in the lattice can be considered to correspond to wave vectors within the first Brillouin zone.

Measurement of crystal-lattice parameters

I should now like to mention very briefly some experimental techniques that can be used to study crystal structure. We would like to know the shape and size of the unit cell and the arrangement of atoms therein. X-ray crystallography is a valuable tool for such studies.

Objects with dimensions of the order of the wavelength of visible light can produce visible diffraction effects. An examination of a diffraction pattern, if the wavelength of the light is known, will tell us something about the dimensions of the object producing it.

The spacings of atoms within a crystal are of the order of 1 Å (10^{-10} metres). The wavelength of *visible* light is about 5000 times this and in order to obtain diffraction effects we need to use electromagnetic radiation with a wavelength $\sim$1 Å, that is, X-rays. They are produced by firing high-energy electrons through vacuum onto a heavy-metal anode. A crystal is placed in the path of the resulting radiation. The electrons in the crystal scatter the X-rays and the crystal acts as a diffraction grating. The diffraction patterns produced are recorded on a suitable photographic film and afterwards analysed. The analysis can yield a map of the electron density throughout the crystal. The heavier atoms with the most electrons scatter most efficiently. Positions of hydrogen atoms within crystals are difficult to determine with any degree of accuracy.

The best way to study the positions of light elements is by neutron diffraction. Low-energy neutrons fired at a crystal are scattered by atomic nuclei. They will suffer diffraction effects if the wavelength associated with the neutron is of the order of 1 Å. This happens when the energy which equals $p^2/2m = h^2/2m\lambda^2 \sim 0.1$ eV. The light nuclei can produce relatively strong scattering as this does not vary in any regular way with atomic number. In non-magnetic crystals the interaction of the neutrons is only with the nuclei. In magnetic crystals there is an interaction between the magnetic moment of the neutron and the electronic moment.

The surfaces of crystals can be studied by observation of the diffraction of electron beams. Electrons interact so strongly with the atoms that the beam will not penetrate very far.

Diffraction methods can supply a great deal of crystallographic information but the analysis of the results is often extremely complicated. Indeed, X-ray crystallography has become a science unto itself.

Diffraction theory

I do not wish to attempt any discussion of the theory of diffraction except the little that is sufficient to indicate a physical interpretation of those mystical entities – reciprocal lattice vectors.

The Bragg condition for diffraction (an oversimplification) states that diffraction will occur from parallel planes of atoms if part of a plane wave (corresponding for instance to X-rays or electrons) reflected from one plane is in phase with the part of the wave reflected from the next plane. Detailed behaviour of the scattering atoms is ignored. For the two wavefronts to be in phase, MNO must be an integer number of wavelengths (see Fig. 1.18). This type of diagram will probably be familiar from a study of light optics.

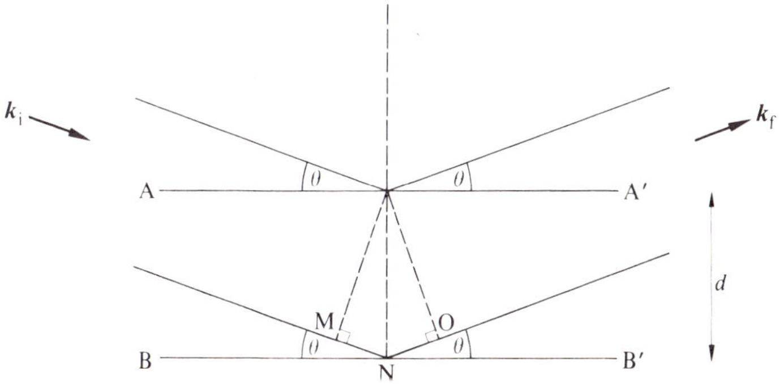

Fig. 1.18. Demonstrates the Bragg condition for diffraction.

Diffraction will then take place if MNO = $n\lambda$ where λ is the wavelength of the radiation.

$$2d \sin \theta = n\lambda.$$

Let us consider that $n = 1$, i.e., that we have the condition for first-order diffraction.

$$2d \sin \theta = \lambda.$$

On diffraction part of the energy in the incident beam, with wavevector $\boldsymbol{k}_i$, will be scattered into a beam marked $\boldsymbol{k}_f$ in Fig. 1.18. How does $\boldsymbol{k}_i$ differ from $\boldsymbol{k}_f$? It is only the component of $\boldsymbol{k}_i$ perpendicular to the planes AA′ and BB′ which is changed. $\boldsymbol{k}_{\parallel}$ remains constant but $\boldsymbol{k}_{\perp}$ changes direction on diffraction. The change of wavevector upon diffraction is perpendicular to the planes and the magnitude is

$$2k \sin\theta = 2\frac{2\pi}{\lambda}\sin\theta = \frac{2\pi}{d}. \qquad \left(k = \frac{2\pi}{\lambda}, \text{ by definition.}\right)$$

Now, planes in a crystal are represented by Miller indices *h, k,* and *l*. The plane (*hkl*) is the plane which has intersections on the three crystal axes ($\boldsymbol{a}_1, \boldsymbol{a}_2$, and $\boldsymbol{a}_3$) proportional to $a_1/h, a_2/k$, and a_3/l. h, k, and l are always reduced to the lowest integers.† For example, the plane with intersections $\frac{1}{2}a_1, a_2$, and a_3 has $h{:}k{:}l = 2{:}1{:}1$. The plane is (211).

If the planes AA′ and BB′ in the diagram are the (*hkl*) planes, then

$$\left(\frac{1}{h}\boldsymbol{a}_1 - \frac{1}{k}\boldsymbol{a}_2\right) \quad \text{and} \quad \left(\frac{1}{k}\boldsymbol{a}_2 - \frac{1}{l}\boldsymbol{a}_3\right)$$

lie in one of the diffraction planes. Now we need a vector which is perpendicular to these diffraction planes and has a length equal to $2\pi/d$ to represent the change in $\boldsymbol{k}$. Try the reciprocal lattice vector $h\boldsymbol{b}_1 + k\boldsymbol{b}_2 + l\boldsymbol{b}_3$.

$$\left(\frac{1}{h}\boldsymbol{a}_1 - \frac{1}{k}\boldsymbol{a}_2\right) \cdot (h\boldsymbol{b}_1 + k\boldsymbol{b}_2 + l\boldsymbol{b}_3) = 2\pi - 2\pi = 0$$

and

$$\left(\frac{1}{k}\boldsymbol{a}_2 - \frac{1}{l}\boldsymbol{a}_3\right) \cdot (h\boldsymbol{b}_1 + k\boldsymbol{b}_2 + l\boldsymbol{b}_3) = 2\pi - 2\pi = 0.$$

The reciprocal lattice vector is perpendicular to the planes.

$$\text{Distance between planes } d = \frac{1}{h}\boldsymbol{a}_1 \cdot \frac{(h\boldsymbol{b}_1 + k\boldsymbol{b}_2 + l\boldsymbol{b}_3)}{|(h\boldsymbol{b}_1 + k\boldsymbol{b}_2 + l\boldsymbol{b}_3)|}$$

$$= \frac{2\pi}{\text{length of reciprocal lattice vector}}$$

$$\text{Length of reciprocal lattice vector} = \frac{2\pi}{d} = \text{change in } \boldsymbol{k}.$$

†The 'diffraction indices' are not always reduced to the lowest possible integers. Because crystallographers prefer to always think of first-order diffraction, they may refer to first-order diffraction from the (222) planes rather than to second-order diffraction from the (111) planes.

Examples

1. 1971 Pure Physics Option I

A two-dimensional direct lattice is formed from a repetition of the points ABCD shown in Fig. 1.19 in which AB = CD = 3 Å and AD = BC = 5 Å; angle BAD = 60°. Draw a small area of the direct lattice and next to it, at the correct relative orientation, the reciprocal lattice. Calculate the lengths of the basis vectors of this reciprocal lattice.

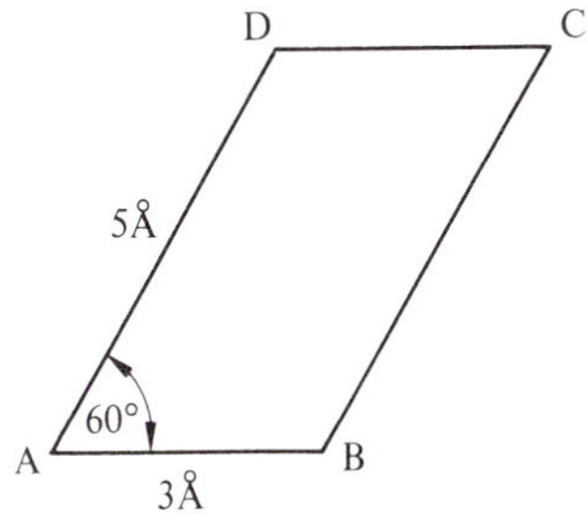

Fig. 1.19.

Answer

Let $\boldsymbol{a}_1$ be the lattice vector corresponding to AB and let $\boldsymbol{a}_2$ be the lattice vector corresponding to AD.

$$\boldsymbol{b}_1 \cdot \boldsymbol{a}_1 = 2\pi \qquad \boldsymbol{b}_1 \cdot \boldsymbol{a}_2 = 0.$$

Therefore $\boldsymbol{b}_1$ is perpendicular to $\boldsymbol{a}_2$ and $b_1 = 2\pi/(3 \cos 30°)\,\text{Å}^{-1} = 4\pi/(3\sqrt{3})\,\text{Å}^{-1}$. In a similar way $\boldsymbol{b}_2$ is perpendicular to $\boldsymbol{a}_1$ and $\boldsymbol{b}_2 = 2\pi/(5 \cos 30°)\,\text{Å}^{-1} = 4\pi/(5\sqrt{3})\,\text{Å}^{-1}$.

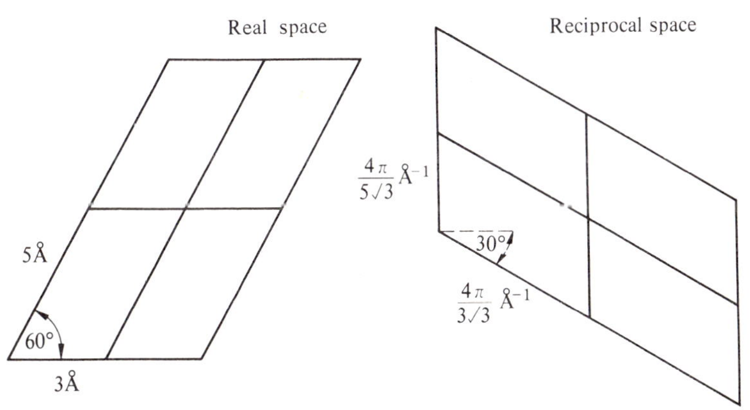

Fig. 1.20.

2. 1972 Pure Physics Option I

Show that the length of the shortest reciprocal lattice vector G in any direction in reciprocal space is related to the smallest separation d between planes of atoms perpendicular to that direction in the crystal structure by $|G| = 2\pi/d$.

Answer

See the last section of chapter 1 for a way to attempt a mathematical proof of this. If a lot of maths worries you try the following approach.

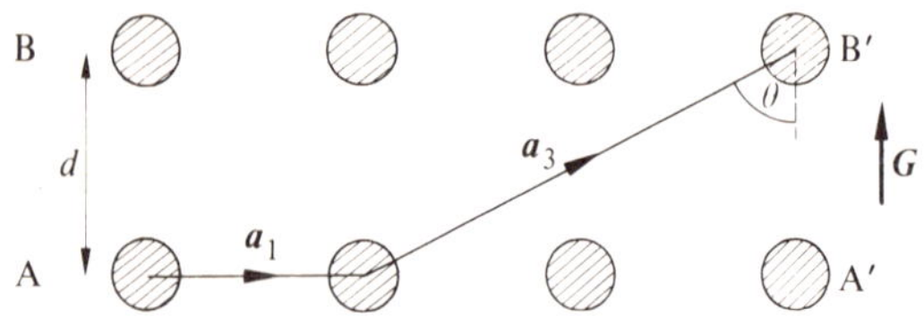

Fig. 1.21.

The diagram shows two planes (AA′ and BB′) of atoms perpendicular to G. Let us say that G is one of a set of fundamental reciprocal lattice vectors corresponding to a set of basis lattice vectors a_1, a_2, and a_3. Two of the a's (say a_1 and a_2) must be perpendicular to G, i.e., in the plane (see p. 11). In order that the a's can cover the *whole* crystal, a_3 must go from one plane to a *nearest-neighbour* plane.

Then

$$G \cdot a_3 = 2\pi \quad \text{(by definition)}$$

becomes

$$|G|\, a_3 \cos\theta = 2\pi$$

$$|G|\, d = 2\pi \quad \text{or} \quad |G| = 2\pi/d.$$

Questions

1. *Bristol, 1974*

Calculate the parameters of the lattice which is reciprocal to the two-dimensional net made up of equilateral triangles of side a. Sketch the first two Brillouin zones.

2. *Manchester, 1973*

Define the terms *lattice point, unit cell*, and *basis* as applied to crystalline solids, illustrating your answers by using sodium chloride as an example.

Copper (atomic weight = 64) crystallizes into a f.c.c. structure with a density of 8930 kg m^{-3}. Calculate the nearest-neighbour spacing of copper atoms.

[Atomic mass unit = 1.66 10^{-27} kg.]

3. *Manchester, 1971*

A metal which has a body-centred-cubic structure has a density of 10 g cm^{-3}. If the atomic mass is 50 a.m.u., what is the separation between nearest-neighbour atoms?

[1 a.m.u. = 1.66 10^{-27} kg.]

4. *Manchester, 1971*

Sketch the potential-energy curve for two neutral argon atoms as a function of their separation. Explain the physical reasons for the form of the curve but do not attempt a mathematical derivation of the detailed form of the potential. How does the curve differ for ionic binding?

Explain how a potential-energy curve can be used to estimate:

(a) the density of a substance at 0 K,

(b) the latent heat of vaporization,

(c) the coefficient of thermal expansion.

5. *Manchester, 1974*

Briefly discuss the origin of the interatomic forces which exist between two atoms of an inert element.

The bulk modulus of a solid can be expressed in the form:

$$K = \frac{n}{18a_0}\left(\frac{d^2 V}{dr^2}\right)_{r=a_0}$$

where n is the coordination number and a_0 is the equilibrium molecular spacing. For simple molecular solids, $V(r)$, the intermolecular potential, can be approximated by the Lennard-Jones equation:

$$V(r) = \epsilon\left[\left(\frac{a_0}{r}\right)^{12} - 2\left(\frac{a_0}{r}\right)^{6}\right].$$

Given that the bulk modulus of solid carbon tetrachloride is 1.1 10^9 J m^{-2} and that the equilibrium separation of its molecules is 5.4 10^{-10} m estimate its latent heat of sublimation.

References

1. C. Kittel. *Introduction to Solid State Physics*, chapter 3, Wiley, 1971.
2. L. Pauling. *The Chemical Bond*. Cornell UP and Oxford UP, 1967.

2. Thermal Properties of Insulators

You might be puzzled by the title of this chapter. Why should we treat insulators in a different way to conductors? The reasons are as follows. The two topics which will be considered here are specific heat and thermal conductivity. The specific heat is the amount of energy that must be supplied to a mole† of the solid in order to raise its temperature by one degree. Some thermal energy is associated with lattice vibrations – the higher the temperature, the greater the amplitude of the vibrations. The specific heat of an insulator depends only‡ on the energy needed to increase the lattice vibrations – it is the lattice specific heat. In a metal there are electrons that are free (or nearly free – see chapters 4 and 5) to move around in the solid: there is an electronic contribution to the specific heat due to the increase in kinetic energy of these electrons when the temperature is raised. The specific heat of a metal is therefore the lattice specific heat *plus* the electronic specific heat.

The free electrons in metals will transfer heat from a 'hot' region of the solid to a 'cold' region. The electrons in the hot region have greater kinetic energy than those in the cold region and as they diffuse towards the cold region they gradually lose the excess energy and reach thermal equilibrium with their new surroundings. Electrons arriving in the hot region from the cold region gain energy from their surroundings. Thus there is a constant transfer of heat energy at a rate which one expects to depend on the specific heat of the electron 'gas' and the rate of

†A mole is the atomic weight of an element or the molecular weight of a compound in grams. The specific heat can also be defined in terms of unit mass.
‡There are other contributions to the specific heat which I shall not deal with here, such as the anomalies associated with magnetic crystals. These contributions occur over a small temperature range – usually at low temperatures – and give rise to peaks on a graph of specific heat against temperature.

diffusion. This contribution to the thermal conductivity occurs *in addition* to the lattice conductivity of an insulator.

Specific heats

Before we attempt calculations on the specific heats of insulators we make an assumption about the nature of the lattice vibrations. In chapter 1 the atoms were treated as fixed at definite positions in the lattice. These are equilibrium positions about which the atoms vibrate. The assumption is made that the restoring force is proportional to the displacement: such forces are called 'harmonic' and they satisfy Hooke's Law. Suppose all the atoms in the solid are fixed at their equilibrium positions except one which belongs at a position r_0. Then if this atom is displaced to a position r the potential energy is assumed to be of the form $V = f(r - r_0)^2$. The atom will execute simple harmonic oscillations of the type $A \cos \omega t$ about r_0. Note that

$$(r - r_0)^2 = (x - x_0)^2 + (y - y_0)^2 + (z - z_0)^2$$

and so the atom is, in effect, able to oscillate independently in the three Cartesian directions. The equation of motion is only separable in this way for the harmonic potential. The atom 'sees' a potential of the form shown in Fig. 2.1. Of course, in a real crystal the potential will not necessarily be so simple but this should be a reasonable approximation for small displacements.

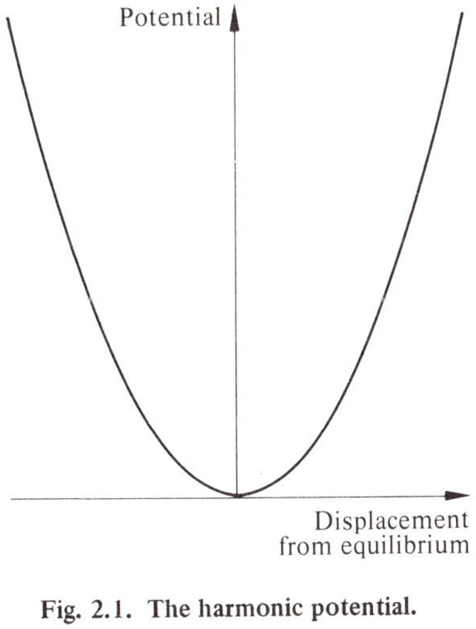

Fig. 2.1. The harmonic potential.

The classical model

Can we guess the specific heat of a solid? The specific heat of an ideal monatomic gas is $\frac{3}{2}k_B$ per atom per degree; the kinetic energy of an atom of the gas at temperature T is $\frac{3}{2}k_BT$. T is the absolute temperature and k_B is Boltzmann's constant. The gas atom has three degrees of freedom. It can have kinetic energy independently in three directions at right angles. We can therefore say that energy $\frac{1}{2}k_BT$ is associated with each degree of freedom. This is an average energy which any atom will have if its energy is averaged over a long enough time or if we consider a large number of atoms at a fixed time.

Consider a crystal of N atoms (where N is a large number!). How many degrees of freedom are associated with these N atoms? Each atom has three independent directions in which it can move, but with each direction there are associated two degrees of freedom because energy can appear as kinetic energy or as potential energy. Therefore the number of degrees of freedom in the crystal is $2 \times 3 \times N = 6N$. The average energy is $6N \times \frac{1}{2}k_BT$ and the specific heat $= 3Nk_B$ or $3R$ where R is the gas constant. This is an argument often given in books and it is certainly an easy way of remembering the answer. Why should a degree of freedom be associated with potential energy as well as with kinetic energy? It can be shown that the expression for the average energy of a one-dimensional simple harmonic oscillator (see eq. (2.1) following) can be written as the sum of two terms – one associated with kinetic energy and the other with potential energy – and that each of these terms is equal to $\frac{1}{2}k_BT$.

I feel that it is better to avoid the degree-of-freedom argument and to calculate the specific heat using the method outlined in the next few paragraphs. Equation (2.1) will be found to be fairly straightforward and an understanding of this expression is necessary for more detailed treatment of specific heats.

Classical statistical mechanics gives the probability that a particle in a system (in thermal equilibrium at an absolute temperature T) shall have an energy between ϵ and $\epsilon + d\epsilon$ as

$$\sim \text{(the number of states between } \epsilon \text{ and}$$

$$\epsilon + d\epsilon = D(\epsilon)d\epsilon) \times \exp(-\epsilon/k_BT).$$

The calculations that yield the term $\exp(-\epsilon/k_BT)$ and the expression for $D(\epsilon)$ involve a formal mathematical procedure of dividing the energy into a set of discrete allowed values and then assuming that the separation of these states is infinitely small. For a 'classical' system, in which the energy forms a continuum, this 'quantization' could have no physical significance. The energy levels of a monatomic gas and of a vibrating atom in a crystal are quantized. Classical statistics will apply in these cases provided that the separation between the allowed energy levels is small enough. The conditions under which this is true will be explained later.

Classical statistics applied to the ideal monatomic gas mentioned earlier gives

$$D(\epsilon)\mathrm{d}\epsilon \propto \epsilon^{1/2}\,\mathrm{d}\epsilon.$$

The reason why $D(\epsilon)\mathrm{d}\epsilon$ is not just proportional to $\mathrm{d}\epsilon$ is because the energy levels are degenerate, i.e., the atom can be in more than one state with an energy ϵ. Provided $|\boldsymbol{p}|^2 = 2m\epsilon$, any value of the momentum $\boldsymbol{p}$ corresponds to a state with energy ϵ. The average energy of an atom in the gas is

$$\langle\epsilon\rangle = \frac{\int_0^\infty \epsilon n_{\epsilon\to\epsilon+\mathrm{d}\epsilon} \exp(-\epsilon/k_\mathrm{B}T)\,\mathrm{d}\epsilon}{\int_0^\infty n_{\epsilon\to\epsilon+\mathrm{d}\epsilon} \exp(-\epsilon/k_\mathrm{B}T)\,\mathrm{d}\epsilon}$$

$$= \frac{\int_0^\infty \epsilon\epsilon^{1/2} \exp(-\epsilon/k_\mathrm{B}T)\,\mathrm{d}\epsilon}{\int_0^\infty \epsilon^{1/2} \exp(-\epsilon/k_\mathrm{B}T)\,\mathrm{d}\epsilon} = \tfrac{3}{2} k_\mathrm{B}T.$$

The denominator is there for normalization purposes. The total probability of being in any state must be unity.

Now consider a system of one-dimensional simple harmonic oscillators which interact sufficiently to reach thermal equilibrium but not sufficiently to be able to change places. The energy levels of a given oscillator are not degenerate in this case. One value of ϵ corresponds to one particular amplitude of the vibration of this oscillator about its equilibrium position. Here

$$D(\epsilon)\mathrm{d}\epsilon \propto \mathrm{d}\epsilon \quad \text{and} \quad \langle\epsilon\rangle = \frac{\int_0^\infty \epsilon \exp(-\epsilon/k_\mathrm{B}T)\,\mathrm{d}\epsilon}{\int_0^\infty \exp(-\epsilon/k_\mathrm{B}T)\,\mathrm{d}\epsilon} = k_\mathrm{B}T. \tag{2.1}$$

As simple harmonic motion in one Cartesian direction is independent of such motion in the other two directions, it might be expected that $\langle\epsilon\rangle$ for a three-dimensional oscillator would be simply $3k_\mathrm{B}T$. This is indeed the case. In three dimensions the energy levels are degenerate in such a way that

$$D(\epsilon)\mathrm{d}\epsilon \propto \epsilon^2\,\mathrm{d}\epsilon^\dagger \quad \text{and} \quad \langle\epsilon\rangle = 3k_\mathrm{B}T.$$

†The number of ways of arranging N_e bits of energy ($\epsilon = N_\mathrm{e}\delta\epsilon$) between three states (three dimensions) is equal to the number of ways of arranging N_e identical objects and two identical divisions

$$= \frac{(N_\mathrm{e}+2)!}{N_\mathrm{e}!2!} = \frac{(N_\mathrm{e}+2)(N_\mathrm{e}+1)}{2} = \frac{N_\mathrm{e}^2}{2} \quad \text{when} \quad N_\mathrm{e}\to\infty.$$

This is proportional to ϵ^2.

If we assume that each atom in a crystal vibrates independently of all the other atoms and behaves as a classical three-dimensional simple harmonic oscillator, then the average energy associated with a crystal of N atoms should be $N\langle \epsilon \rangle = N3k_BT$. A monatomic crystal with $N = N_A$, Avogadro's number, contains one mole of the element.

$$\text{Energy} = 3RT \text{ per mole;}$$

$$\text{specific heat} = \frac{\partial E}{\partial t} = 3R \text{ mole}^{-1} \text{ degree}^{-1} \sim 6 \text{ cal mole}^{-1} \text{ K}^{-1}.$$

R is the general gas constant and is equal to $N_A k_B$. Strictly, this is the specific heat at constant volume C_v. (The specific heat at constant pressure C_p is measured experimentally. C_v can be obtained from C_p plus compressibility and expansivity coefficients. $C_p - C_v$ is fairly small for solids.)

This is a simple way of arriving at an estimate of the specific heat of a solid, but is it correct? At about room temperature and above this result is borne out (approximately) by experimental measurements and is, in fact, equivalent to the law of Dulong and Petit. This law states that: The product of the atomic weight and the specific heat (per unit mass) is the same for all elementary solid substances.

However, classical statistics suggest that this estimate of specific heat should apply at all temperatures even down to absolute zero. Experimental measurements show that the specific heat of a solid drops towards zero as T approaches 0 K.

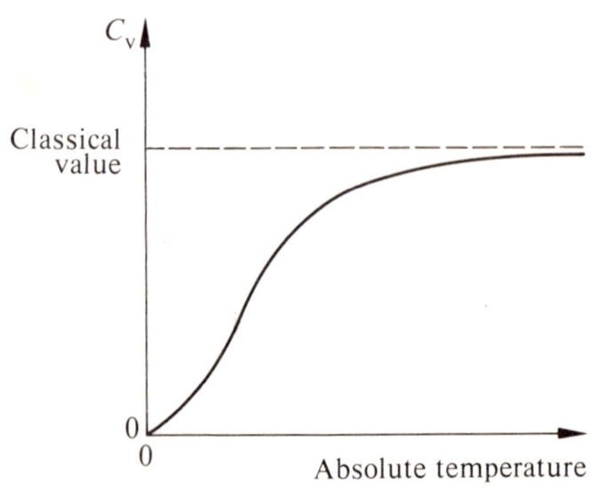

Fig. 2.2. Sketch of the behaviour of the specific heat of a solid at low temperatures.

The Einstein model

Einstein was the first person to produce an explanation of this phenomenon. He treated each atom in a solid as a harmonic oscillator moving independently of all the other atoms – as for the classical calculation – *but* he made use of Planck's quantum postulate. Planck had said that the energy levels of electromagnetic oscillations are quantized. Following this Einstein said that the energy levels of an atom vibrating in one dimension do not form a continuum. Instead, an oscillator with angular frequency $\omega(2\pi \times \text{frequency})$ has allowed energy levels separated by $\hbar\omega$ (see Fig. 2.3(b)). We can write

$$\epsilon_n = n\hbar\omega \quad \text{where} \quad n = 0,1,2,\ldots$$

Then the integrals of classical statistical mechanics must be replaced by summations.

$$\text{Average energy of one-dimensional oscillator} = \langle \epsilon \rangle$$

$$= \sum_{\substack{\text{all allowed}\\ \text{values}}} \epsilon_n \exp(-\epsilon_n/k_B T) \Big/ \sum_{\substack{\text{all allowed}\\ \text{values}}} \exp(-\epsilon_n/k_B T)$$

$$= \sum_{n=0}^{\infty} n\hbar\omega \exp(-n\hbar\omega/k_B T) \Big/ \sum_{n=0}^{\infty} \exp(-n\hbar\omega/k_B T).$$

This expression looks difficult to evaluate but in fact it is easy if one knows the right way to set about it. First consider the term $\Sigma \exp(-n\hbar\omega/k_B T)$ (call it S). This is a geometrical progression of the form $a + ar + ar^2 + \ldots$ and the sum is $a/(1-r)$. As in this case $a = 1$ and r is $\exp(-\hbar\omega/k_B T)$ the sum S is $1/[1-\exp(-\hbar\omega/k_B T)]$. Then we can differentiate S with respect to $1/k_B T$ and we find

$$\frac{\partial S}{\partial(1/k_B T)} = -\sum_{n=0}^{\infty} n\hbar\omega \exp(-n\hbar\omega/k_B T) = \frac{-\hbar\omega \exp(-\hbar\omega/k_B T)}{[1-\exp(-\hbar\omega/k_B T)]^2}.$$

This little trick enables us to do the summation in the numerator of the expression for $\langle \epsilon \rangle$.

Then

$$\langle \epsilon \rangle = \frac{\hbar\omega \exp(-\hbar\omega/k_B T)}{[1-\exp(-\hbar\omega/k_B T)]^2} \Big/ \frac{1}{1-\exp(-\hbar\omega/k_B T)}$$

$$= \frac{\hbar\omega \exp(-\hbar\omega/k_B T)}{1-\exp(-\hbar\omega/k_B T)}$$

or

$$\frac{\hbar\omega}{\exp(\hbar\omega/k_B T) - 1}.$$

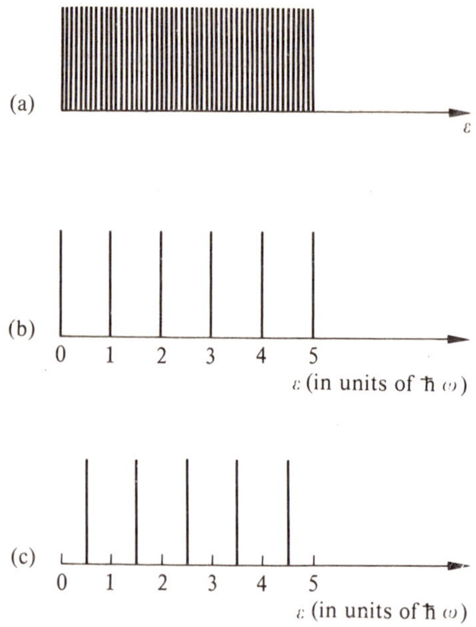

Fig. 2.3. Energy levels of a one-dimensional harmonic oscillator of angular frequency ω. (a) Classical model. (b) Einstein's model. (c) Actual levels from solution of the Schrödinger wave equation.

For a solid element containing N atoms and therefore effectively $3N$ one-dimensional oscillators we have

$$E = 3N \frac{\hbar\omega}{\exp(\hbar\omega/k_B T) - 1}.$$

If $\hbar\omega$ is very small in comparison with $k_B T$ the exponential can be written

$$\exp(\hbar\omega/k_B T) = 1 + \frac{\hbar\omega}{k_B T} + \ldots$$

and

$$E = 3N \frac{\hbar\omega}{1 + (\hbar\omega/k_B T) - 1} = 3Nk_B T \text{ which is the classical result.}$$

The expression for the specific heat is

$$C_v = \frac{\partial E}{\partial T} = 3N \frac{\partial}{\partial T}\left(\frac{\hbar\omega}{\exp(\hbar\omega/k_B T) - 1}\right)$$

$$= 3Nk_B \left(\frac{\hbar\omega}{k_B T}\right)^2 \frac{\exp(\hbar\omega/k_B T)}{[\exp(\hbar\omega/k_B T) - 1]^2}.$$

Again, if $\hbar\omega$ is small in comparison with $k_B T$

$$C_v = 3Nk_B \left(\frac{\hbar\omega}{k_B T}\right)^2 \frac{1}{(\hbar\omega/k_B T)^2} = 3Nk_B.$$

This is the classical value for specific heat. If the temperature is high, so many of the discrete levels will be occupied that the total energy (and hence the specific heat) of the system will not be appreciably different from what it would be if the allowed levels formed a continuum. However, at low temperatures only one or two of the allowed levels will be occupied to any extent and the energy will be strongly dependent on the exact spacing of the levels.

What happens to C_v as $T \to 0$ K?

$$C_v = 3Nk_B \left(\frac{\hbar\omega}{k_B T}\right)^2 \frac{\exp(\hbar\omega/k_B T)}{[\exp(\hbar\omega/k_B T) - 1]^2}$$

$$\to 3Nk_B \left(\frac{\hbar\omega}{k_B T}\right)^2 \frac{\exp(\hbar\omega/k_B T)}{[\exp(\hbar\omega/k_B T)]^2}$$

$$= 3Nk_B \left(\frac{\hbar\omega}{k_B T}\right)^2 \Big/ \exp(\hbar\omega/k_B T).$$

As $x \to \infty$, $\exp(x)$ goes to infinity more rapidly than any power of x, so we have $C_v \to 0$.

This is the type of behaviour observed experimentally and Einstein's method yielded theoretical curves which fitted experimental curves quite well. The comparison was not perfect. Experiments showed that at low temperatures C_v was proportional to T^3, whereas the calculations above seem to indicate an exponential dependence. Still, one doesn't expect perfect agreement from such a simple model, and Einstein had succeeded in showing that the behaviour of C_v at low temperatures was related to some quantization of the allowed energy levels in the solid. The atoms in a solid do not all vibrate independently at a fixed frequency. Vibrational waves of various frequencies travel through the solid.

Before we leave the Einstein model I should just like to mention one point. Planck's postulates when applied to the quantization of the energy levels of a simple harmonic oscillator were essentially correct. The separation between allowed levels is $\hbar\omega$ but wave-mechanical calculations show that the actual energy levels are given by

$$\epsilon_n = (n + \tfrac{1}{2})\hbar\omega \quad \text{where} \quad n = 0,1,2,\ldots$$

This means that even at absolute zero of temperature an oscillator will have a finite energy + ½ħω. This is known as the zero-point energy. If we go through the calculation again we find that

$$\langle \epsilon \rangle = \frac{\sum_{n=0}^{\infty} (n + ½)\hbar\omega \exp[-(n + ½)\hbar\omega/k_B T]}{\sum_{n=0}^{\infty} \exp[-(n + ½)\hbar\omega/k_B T]}$$

$$= \frac{\sum_{n=0}^{\infty} n\hbar\omega \exp(-n\hbar\omega/k_B T)\exp(-½\hbar\omega/k_B T)}{\sum_{n=0}^{\infty} \exp(-n\hbar\omega/k_B T)\exp(-½\hbar\omega/k_B T)}$$

$$+ \frac{\sum_{n=0}^{\infty} \hbar\omega \exp[-(n + ½)\hbar\omega/k_B T]}{\sum_{n=0}^{\infty} \exp[-(n + ½)\hbar\omega/k_B T]}$$

$$= \frac{\sum n\hbar\omega \exp(-n\hbar\omega/k_B T)}{\sum \exp(-n\hbar\omega/k_B T)} + ½\hbar\omega.$$

The result of using this new set of energy levels is to obtain an extra, temperature-independent contribution to the total energy of the solid. Because this term does not depend on temperature it can have no effect on C_V.

The Debye model

An unrealistic aspect of the Einstein model is the restriction on the frequency of the oscillations. The Debye model is the result of an attempt to include vibrational waves of different frequencies. The solid is treated as an elastic continuum. Elastic waves have the property that their wavelength and frequency are related by the simple expression

$$\lambda \times \frac{\omega}{2\pi} = \text{velocity of wave } (v).$$

There are two kinds of vibrational waves. Longitudinal waves are such that the direction of oscillation is the same as the direction of propagation of the wave. For transverse waves the oscillation is at right angles to the direction of propagation. (See Fig. 2.4(a).) There is no reason why the velocity of the longitudinal waves should be the same as that of the transverse waves but I shall assume they are the same (v).

The main predictions of the Debye theory are not affected by this simplification.

In order to do the statistical calculations for the thermal energy of the Debye solid we need to know the number of allowed vibrational states with angular frequencies between ω and $\omega + \mathrm{d}\omega$. $\omega = v(2\pi/\lambda)$ and the term $2\pi/\lambda$ is the magnitude k of the wave vector $\boldsymbol{k}$ which points in the direction of propagation of the wave. Now, suppose we consider a volume of this continuum as shown in Fig. 2.4(b). It is bounded by the planes $x = 0, x = 1, y = 0, y = 1, z = 0$, and $z = 1$. If the boundary planes are 'held stationary' then the wavelength of any vibration must be such that an integral number of half-wavelengths will fit in between $x = 0$ and $x = 1$ (Fig. 2.4(c)). Similar conditions hold for the other two Cartesian directions. If λ_x is the projection of the wavelength λ on the x-axis, etc., then

$$\begin{aligned}\lambda^2 &= \lambda_x^2 + \lambda_y^2 + \lambda_z^2 \\ &= \left(\frac{2l}{n_x}\right)^2 + \left(\frac{2l}{n_y}\right)^2 + \left(\frac{2l}{n_z}\right)^2\end{aligned}$$

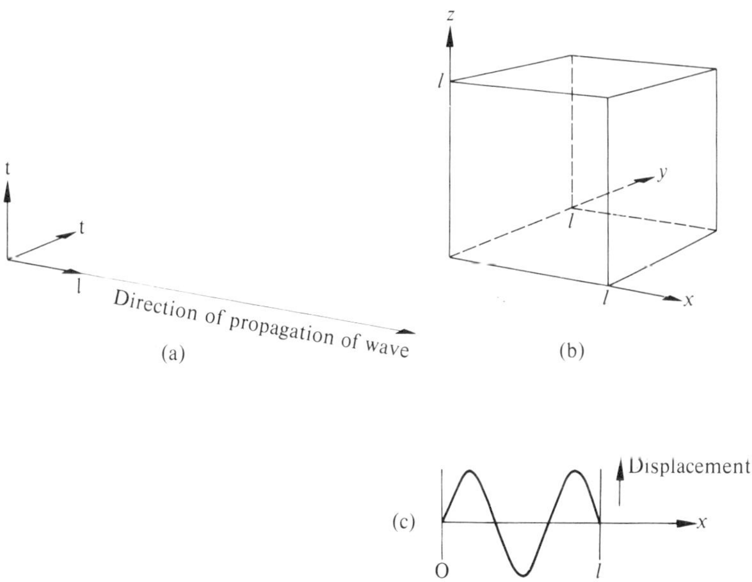

Fig. 2.4. (a) Longitudinal and transverse waves. Thick arrows show directions of displacement for longitudinal waves (l) and for transverse waves (t). (b) A volume of the elastic continuum. For simplicity a cube is considered. (c) A possible value for $\lambda(= 2l/3)$. The wave is shown transverse for clarity.

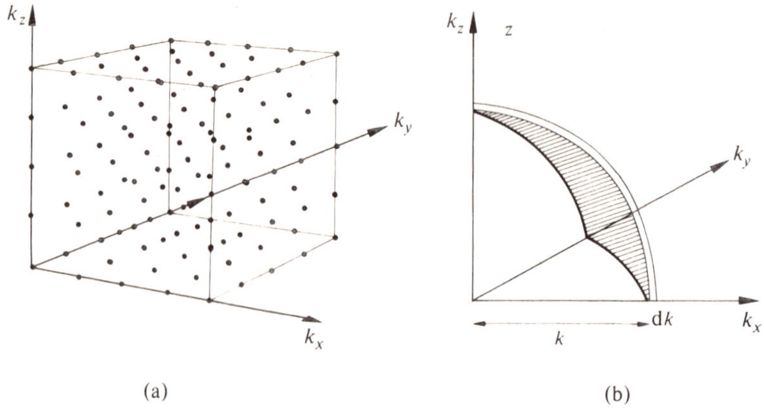

Fig. 2.5. (a) Grid of points in k-space representing allowed values of k. (b) Octants of spheres with radii k and k + dk.

where n_x, n_y, and n_z are any integers. Or

$$\left(\frac{2\pi}{k}\right)^2 = \left(\frac{2\pi}{k_x}\right)^2 + \left(\frac{2\pi}{k_y}\right)^2 + \left(\frac{2\pi}{k_z}\right)^2$$

where $k_x = \pi n_x/l$, etc.

A plot of the allowed values of $\boldsymbol{k}$ in reciprocal space or k-space is a grid of points with x-values of $\pi n_x/l$ $(n_x \geqslant 0)$, y-values of $\pi n_y/l$, and z-values of $\pi n_z/l$ as shown in Fig. 2.5(a). We need only positive integers to label all the standing waves. Each point occupies a volume $(\pi/l)^3$. Suppose that the volume $V = l^3$ gets larger and larger so that the points get closer together. As $V \to \infty$ the number of allowed states between k and k + dk becomes simply the volume in k-space corresponding to $|\boldsymbol{k}|$ between k and k + dk divided by π^3/V. (Fig. 2.5(b).) This is

$$\tfrac{1}{8}\ (\text{positive octant}) \times 4\pi k^2\,\mathrm{d}k \times \frac{V}{\pi^3} \times 3.$$

The factor of three appears because each point corresponds to one longitudinal and two transverse modes. With $\omega = vk$

$$D(\omega)\mathrm{d}\omega = \tfrac{1}{8}4\pi\frac{\omega^2}{v^2}\ \frac{\mathrm{d}\omega}{v}\ \frac{V}{\pi^3}3 = \frac{3V}{2\pi^2 v^3}\,\omega^2\,\mathrm{d}\omega.$$

In order to make the number of modes of vibration equal to the number of modes that would occur in the real solid (a concession to the fact that the solid is not really a continuum) Debye used a cut-off frequency such that

$$\int_0^{\omega_D} D(\omega)\mathrm{d}\omega = 3N$$

where N = the number of atoms in the solid. The expression

$$\int_0^{\omega_D} \frac{3V\omega^2}{2\pi^2 v^3}\,\mathrm{d}\omega = 3N$$

gives

$$\omega_D^3 = \frac{6Nv^3\pi^2}{V}\ .$$

The average energy of an oscillator of frequency ω at temperature T was calculated for the Einstein model and is given by

$$\langle\epsilon\rangle = \frac{\hbar\omega}{\exp(\hbar\omega/k_BT) - 1}\ .$$

So the ènergy E of the Debye solid is given by

$$E = \int_{\substack{\text{all}\\ \text{allowed}\\ \text{frequencies}}} \frac{\hbar\omega}{\exp(\hbar\omega/k_BT) - 1} D(\omega)\mathrm{d}\omega$$

$$= \int_0^{\omega_D} \frac{\hbar\omega}{\exp(\hbar\omega/k_BT) - 1}\,\frac{3V\omega^2}{2\pi^2 v^3}\,\mathrm{d}\omega$$

$$= \frac{3V\hbar}{2\pi^2 v^3}\int_0^{\omega_D} \frac{\omega^3}{\exp(\hbar\omega/k_BT) - 1}\,\mathrm{d}\omega. \qquad (2.2)$$

In the integral above replace $\hbar\omega/k_BT$ by x. With x_D corresponding to $\hbar\omega_D/k_BT$ this gives

$$E = \frac{3V\hbar}{2\pi^2 v^3}\,\frac{k_BT}{\hbar}\left(\frac{k_BT}{\hbar}\right)^3 \int_0^{x_D} \frac{x^3}{e^x - 1}\,\mathrm{d}x.$$

We can define the Debye temperature as θ_D where $\hbar\omega_D = k_B\theta_D$.

Then

$$\left(\frac{k_B}{\hbar}\right)^3 = \left(\frac{\omega_D}{\theta_D}\right)^3 = \frac{6N\pi^2 v^3}{V}\,\frac{1}{\theta_D^3}$$

and the energy at temperature T can be written as

$$E = 9N\,k_B T \left(\frac{T}{\theta_D}\right)^3 \int_0^{x_D} \frac{x^3}{e^x - 1}\,dx$$

which does not depend explicitly on the volume and looks more like the expressions we had for the classical and the Einstein models. We can look at the form the integral will take at high and low extremes of temperature. Now x was $\hbar\omega/k_B T$ and if T is so large that $k_B T \gg \hbar\omega$ even when $\omega = \omega_D$ then x is small over the whole range of the integration and

$$\int_0^{x_D} \frac{x^3}{e^x - 1}\,dx \to \int_0^{x_D} \frac{x^3}{x}\,dx = \frac{x_D^3}{3} \quad \text{or} \quad \frac{1}{3}\left(\frac{\theta_D}{T}\right)^3 .$$

So $E = 9Nk_B T(1/3)$ or $3Nk_B T$ which is the classical result. In the other extreme when $k_B T \ll \hbar\omega_D$ then the upper limit of the integration can be replaced by infinity.

$$\int_0^{x} \frac{x^3}{e^x - 1}\,dx \to \int_0^{\infty} \frac{x^3}{e^x - 1}\,dx .$$

The value of this integration is in fact $\pi^4/15$ but the actual numerical value is not too important. The important point is that such an integration is a constant – it will not depend on T, θ_D, or any other of our parameters. Let us write

$$I = \int_0^{\infty} \frac{x^3}{e^x - 1}\,dx.$$

Then

$$E = 9Nk_B T\left(\frac{T}{\theta_D}\right)^3 I .$$

The energy depends on the fourth power of the absolute temperature.

$$C_v = \frac{\partial E}{\partial T} = 36Nk_B I \frac{T^3}{\theta_D^3} .$$

The Debye model predicts that for $T \ll \theta_D$ the specific heat should depend on the cube of the absolute temperature. This is just the type of dependence which is observed experimentally and the Debye model was a great triumph for quantum physics. Strictly, θ_D is determined by the number of atoms per unit volume of the solid and by the velocity of the elastic waves. However, we can relax these conditions and make the theory more flexible by regarding θ_D as an adjustable parameter which can be varied in an attempt to fit the theoretical to the experimental curves. How good is the Debye model?

Theoretical curves using a Debye cut-off frequency ω_D calculated from the elastic constants do not fit the experimental curves exactly over all temperatures although the agreement is good considering the simplicity of the model. The theory suggests that there should be a T^3 dependence at temperatures below about one-tenth of the corresponding θ_D but in fact the T^3 dependence only sets in at much lower temperatures – nearer one-hundredth of θ_D. Below $\theta_D/100$ experimental and theoretical curves can be made to agree and a good fit can also be had with temperatures above about $0.2\theta_D$ provided one uses a different 'effective θ_D' in each range. Between these two temperature ranges the agreement is less good – the effective θ_D varies more rapidly with temperature.

From the definition of θ_D, we would expect the Debye temperature to be greatest for solids in which the vibrational modes have the highest frequencies. For a given restoring force, a light atom will vibrate more rapidly than a heavy one. But, also, a light atom tends to form stronger bonds than a similar heavier atom. One therefore expects, other things being equal, that a light element will have a higher θ_D than a heavy element. The alkali metals show this trend: lithium ~350 K, sodium ~160 K, potassium ~90 K, rubidium ~55 K, caesium ~40 K. Three elements with the diamond structure in Group IV have: carbon ~2000 K, silicon ~640 K, germanium ~370 K. Again the lightest element has the highest θ_D. The high θ_D for diamond is not due just to the light mass but also to the very strong bonds. The Group IV elements with their covalent bonding have much higher Debye temperatures than the alkali metals in the same row of the periodic table.

One would not expect the Debye model to be faultless as the solid cannot be represented by a continuum at frequencies near ω_D. This frequency corresponds to a wavelength of the same order as the atomic spacings. The frequency spectrum (the number of possible vibrational modes per unit frequency range plotted against frequency) of a discrete lattice is not the same as that of a continuum. (See Fig. 2.6.)

Why does the Debye model appear to be good for low temperatures and high temperatures but not so good in between? At low temperatures, where C_V in sensitive to $D(\omega)$ only the low-frequency modes are appreciably excited when the temperature is slightly raised. The spacing of the allowed levels of the high-frequency modes (proportional to $\hbar\omega$) is so great that the probability of the 'oscillator' being in any but its ground state is negligible. The low-frequency

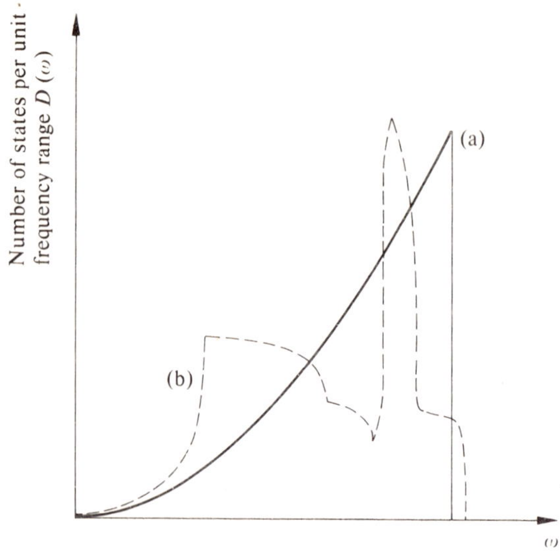

Fig. 2.6. Frequency spectrum – (a) according to the Debye model. (b) Sketch of the type of behaviour to be expected from a real crystal. The areas under the two curves are the same – both equal to the total number of vibrations.

modes which are being excited at these low temperatures have wavelengths which are hundreds or thousands of atomic spacings and it is for just these types of vibration that the lattice can be approximated by a continuum. At higher temperatures the high-frequency modes begin to make a contribution to the specific heat and the Debye model is less good. Above about $0.2\theta_D$ the Debye and Einstein curves are much the same *shape.* This indicates that C_V is not very sensitive to $D(\omega)$ at these temperatures. It does not matter too much that $D(\omega)_{\text{Debye}}$ is not $D(\omega)_{\text{real solid}}$. At still higher temperatures the amount of thermal energy in the crystal is such that many states of even the high-frequency modes have a large probability of being occupied and when $T \gg \theta_D$ the energy levels effectively form a continuum. Here one expects exact frequencies involved to become unimportant. Theoretical and experimental results all tend towards the classical limit as $T \to \infty$.

Can we do better than the Debye model? In order to get a better result we must use the exact frequency spectrum and do the integration of eq. (2.2) numerically. It is possible to get information about the frequency spectrum by studying the inelastic scattering of neutrons. The laws of conservation of energy and momentum (see p. 62) will yield values of $\boldsymbol{k}$ associated with the values of $\hbar\omega$. It is also possible to *calculate* the modes of vibration of a crystal from a knowledge of the crystal structure and the elastic constants. A simple one-dimensional model is presented in the next section and even this is not trivial. The three-dimensional calculations are very much more difficult and would be practically impossible without the help of electronic computers. In practice, the best way to find $D(\omega)$ is from a combination of experimental and theoretical work.

The classical, Einstein, and Debye theories ignore the actual microscopic structure of the solid and each make drastic simplifying assumptions about the way the solid will behave. These theories might therefore be expected to apply to any solid – not necessarily a simple crystalline one. However, the more detailed calculations using a true frequency spectrum are effectively limited by their complexity to simple crystals.

Vibrational waves in a one-dimensional lattice

This simple model is produced to illustrate some of the differences between waves in a discrete lattice and those in a continuum.

In an elastic continuum the displacement due to a wave travelling in the (positive) x-direction can be expressed as the real part of $A \times \exp i(kx - \omega t)$. Here, A is a constant – the amplitude of the wave, k is the magnitude of the wave vector, and ω is the angular frequency. Also, in this particular case, ω and k are related by the expression $\omega = vk$ where v is the velocity of elastic waves in the continuum.

Consider the linear monatomic lattice shown in Fig. 2.7. The wave *displacements cannot be defined at positions* along the x-axis *other than those of the atomic sites* $x = an$ (where n is any integer). A wave here must be expressed in the form $u_n = A \exp\{i(kna - \omega t)\}$ where u_n

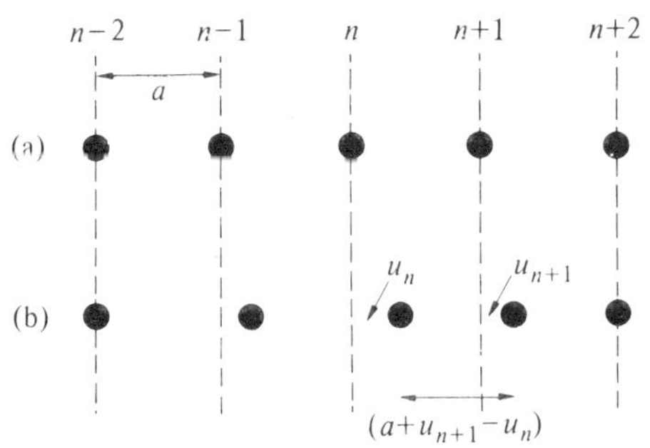

Fig. 2.7. Vibrational waves in a linear monatomic lattice.

is the displacement of the nth atom. The relationship between k and ω it is not as simple as before and must be calculated.

The restoring force will be considered to be harmonic and to exist only between nearest neighbours. In other words the atoms will behave as if they are joined to their nearest neighbours by springs which obey Hooke's law.

Restoring force $= f \times$ extension of the spring. So the force acting on the atom labelled n is

$$f(u_{n+1} - u_n) + f(u_{n-1} - u_n)$$

in the x-direction and the equation of motion of the atom, if it has mass m, is

$$m\frac{\partial^2 u_n}{\partial t^2} = f(u_{n+1} + u_{n-1} - 2u_n). \tag{2.3}$$

Substitution of the expression $u_n = A \exp\{\mathrm{i}(kna - \omega t)\}$ gives

$$\begin{aligned} -m\omega^2 A \exp[\mathrm{i}\{kna - \omega t\}] &= fA \exp[\mathrm{i}\{k(n+1)a - \omega t\}] \\ &+ fA \exp[\mathrm{i}\{k(n-1)a - \omega t\}] - 2fA \exp[\mathrm{i}\{kna - \omega t\}]. \end{aligned} \tag{2.4}$$

$$\begin{aligned} -m\omega^2 &= f\exp(\mathrm{i}ka) + f\exp(-\mathrm{i}ka) - 2f \\ &= f\{\cos ka + \mathrm{i}\sin ka + \cos ka - \mathrm{i}\sin ka - 2\} \\ &= -4f\sin^2\left(\frac{ka}{2}\right) \end{aligned}$$

$$\text{or } \omega = \pm 2\left(\frac{f}{m}\right)^{1/2}\sin\left(\frac{ka}{2}\right).$$

The frequency ω is, by definition, positive. The + or – sign is used, depending on the value of k, to make ω positive.

This expression has a maximum for values of $ka/2$ equal to some (odd) multiple of $\pi/2$ and is zero for values of $ka/2$ equal to multiples of π. A sketch of the curve is shown in Fig. 2.8, together with the curve for an elastic continuum which is the line $\omega = kv$. It can be seen that the discrete lattice behaves very differently from a continuum at values of $ka/2 = \pi/2$, i.e., $k = \pi/a$ or $\lambda = 2\pi/k = 2a$. It is also clear that all the different allowed frequencies can be associated with wave vectors restricted to any range of values which covers $2\pi/a$. A wave that can be described by a particular value of k can also be described by $k + 2\pi/a$, $k + 4\pi/a$, $k + 6\pi/a$, etc. This is a situation which arises simply because the lattice is discrete and we have no means of telling what the wave is doing in the spaces between the atoms.

$$u_n = A \exp\{i(kna - \omega t)\}.$$

If $k \rightarrow k + 2\pi ra$ where r is an integer, then

$$u_n \rightarrow A \exp\left[i\left\{\left(k + \frac{2\pi r}{a}\right) na - \omega t\right\}\right] = A \exp\{i(kna - \omega t)\} \times \exp(i2\pi rn).$$

But $\exp(i2\pi rn) = \cos(2\pi rn) + i \sin(2\pi rn) = 1 + i\,0 = 1$

and, therefore, u_n remains the same.

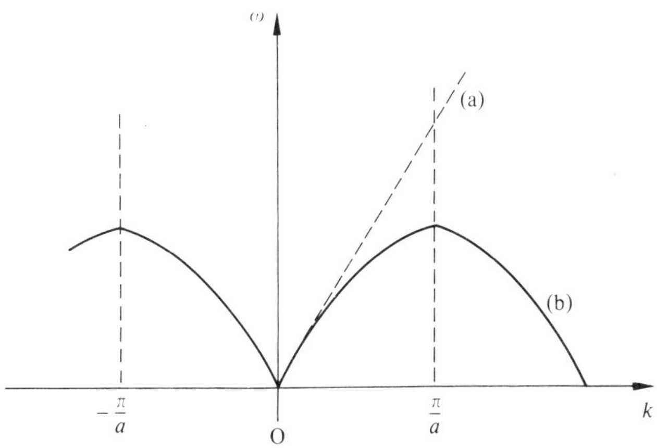

Fig. 2.8. The dispersion relation for (a) a continuum, and (b) a discrete lattice.

The vibrational modes of the one-dimensional monatomic lattice can therefore be represented by a set of k-values (for running waves) over a range of $2\pi/a$. It is usual to take the range $k = -\pi/a \rightarrow k = +\pi/a$. This is the first Brillouin zone of the lattice (see chapter 1). The shortest reciprocal lattice vectors are $-2\pi/a$ and $+2\pi/a$ and the bisectors of these vectors are $-\pi/a$ and $+\pi/a$. In three dimensions we have a similar situation. All waves can be represented by $\boldsymbol{k}$-values within the first Brillouin zone. Near the boundaries of the zone the dispersion relation (ω versus $\boldsymbol{k}$) is unlike that of an elastic continuum.

If we use standing waves instead of running waves k will always be positive. The displacement can be written as

$$u_n = A \sin(kna) \cos(\omega t)$$

if the atom at $n = 0$ is on the boundary and therefore stationary. Replacing k by $k + 2\pi/a$ gives

$$u_n' = A \sin(kna + 2\pi n) \cos(\omega t)$$

which means that $u_n' = u_n$ and the wave is unchanged. *But* replacing k by $-k + 2\pi/a$ gives

$$u_n' = A \sin(-kna + 2\pi n) \cos(\omega t).$$

Sin $(x) = -\sin(-x)$ so $u_n' = -u_n$ for all n. However, if the signs of *all* the displacements change then the vibrational *state* remains the same. The displacements u_n' simply correspond to a wave π radians out of phase with the wave described by the displacements u_n. Hence for standing waves, all k-values can be fitted between $k = 0$ and π/a.

There is no discrepancy here. The total number of allowed modes is the same whether we use travelling waves or standing waves. For standing waves the number of allowed states per given range of k-values is calculated from *fixed boundary conditions.* Suppose that there are $N + 1$ atoms (N is large) in the one-dimensional chain and that the atoms at either end are held fixed. The length of the chain is Na where a is the atomic spacing. An integer number of half-wavelengths must fit into a distance Na so

$$n_1 \frac{\lambda}{2} = Na, \qquad n_1 = 1 \rightarrow N - 1.$$

When $n_1 = N$ all the atoms are stationary. If we allow movement of the chain as a whole we can have $n_1 = 0$ or $n_1 = N$ or both states counted at half-weight.

The allowed values of k are given by

$$k = \frac{2\pi}{\lambda} = \frac{\pi n_1}{Na}, \qquad n_1 = 1 \rightarrow N - 1, \qquad n_1 = 0 \text{ and } N, \text{ at half-weight.}$$

When dealing with running waves we employ *cyclic boundary conditions* (so called because one must imagine a circular chain of atoms). Such boundary conditions can be expressed by the equation

$$u_n(t) = u_{n+N}(t)$$

when there are N atoms in the chain.

$$A \exp\,[\mathrm{i}\{kna - \omega t\}] = A \exp\,[\mathrm{i}\{k(n+N)a - \omega t\}]$$

or

$$\exp(\mathrm{i}kNa) = 1.$$

This means that $kNa = 2\pi n_1$ where n_1 is an integer $0, \pm 1, \pm 2, \ldots$

$$k = \frac{2\pi n_1}{Na}.$$

In order that the total number of modes shall be equal to N, n_1 must take N different values.

$$k = \frac{2\pi n_1}{Na}, \quad n_1 = 0, \pm 1, \pm 2, \ldots \frac{\pm N}{2}.$$

The value of $k = 0$ corresponds to the movement of the chain as a whole. The state with $k = \pi/a$ is the same as the state with $k = -\pi/a$ so if both signs of $n_1 = \pm N/2$ are to be used the states must be taken at half-weight. There are twice as many allowed vibrational states per unit range of k-values for standing waves as for running waves.

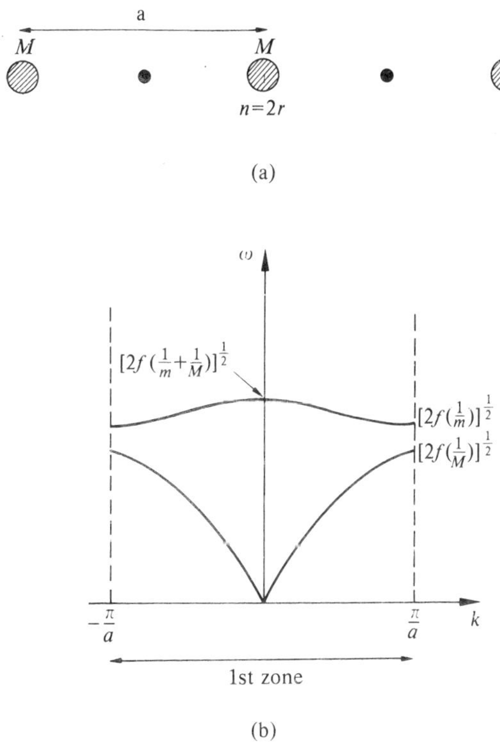

Fig. 2.9. (a) A one-dimensional diatomic lattice. (b) The dispersion relationship for the lattice shown in (a).

The diatomic lattice The one-dimensional lattice with equally spaced atoms such that atoms $n = 0, \pm 2, \pm 4$, etc., have mass M and atoms $n = \pm 1$, ± 3, etc., have mass m is a particular example of a lattice with a basis. The chain of atoms is illustrated in Fig. 2.9(a).

Here we need to define two types of displacements u (corresponding to even n and to odd n) for each allowed wave vector. Solutions will be of the form

$$u_{2r} = A \exp\,[\mathrm{i}\{2kr\,a/2 - \omega t\}]$$

$$u_{2r+1} = B \exp[\mathrm{i}\{2k(r+1)a/2 - \omega t\}]$$

where r is an integer.

An analysis similar to that for the monatomic lattice shows that there is a solution to the equations of motion of the lattice when

$$\omega^2 = f\left(\frac{1}{m} + \frac{1}{M}\right) \pm f\left\{\left(\frac{1}{m} + \frac{1}{M}\right)^2 - \frac{4\sin^2(ka/2)}{mM}\right\}^{1/2}. \qquad (2.5)$$

The first Brillouin zone again extends from $k = -\pi/a$ to $k = +\pi/a$ but here a is the spacing of similar atoms. The nearest-neighbour spacing is $a/2$. For each allowed k-value in the zone there are two allowed frequencies. The dispersion relation is plotted in Fig. 2.9(b). Note that there is a forbidden frequency range between $\omega = (2f/M)^{1/2}$ and $(2f/m)^{1/2}$. The lower-frequency or acoustic branch corresponds to wave motion in which A/B is always positive, i.e., both types of atom move in the same direction. In the higher-frequency or optical branch A/B is never positive. The two types of atoms move in opposite directions.

This high-frequency branch was given the name 'optical branch' because early experimenters on *ionic* crystals showed that it was associated with the absorption of electromagnetic radiation. In ionic crystals the different types of atoms carry positive and negative charges. Such crystals are known to absorb strongly in the infrared at frequencies which correspond to the maxima of the optical branches in these (three-dimensional) crystals. The oscillating electric field of the electromagnetic radiation acts in opposite directions on the positive and negative ions causing them to vibrate in such a way that A/B is negative for a diatomic crystal. A similar set of relationships hold for more complicated crystals.

Thermal conductivity

We have seen that the vibrational modes of a one-dimensional lattice in the harmonic approximation are solutions of eq. (2.3). Let us consider such a 'crystal' at absolute zero. All the allowed modes of vibration have energy $\frac{1}{2}\hbar\omega$ – their zero-point energy. Now suppose that at one end of the chain we add energy to a vibrational mode with frequency ω_1. The smallest amount of energy which we can add to this mode is $\hbar\omega_1$ and this 'packet' of energy will travel through the crystal with a velocity equal to the velocity of the elastic wave in the crystal. This energy packet is called a *phonon.*

Now suppose that we add a phonon of energy $\hbar\omega_2$ to a mode with frequency ω_2. This second phonon will move through the crystal just as it would if the first phonon were not there. There is no interaction between the different vibrational waves. This can be seen mathematically from eq. (2.4). Because only the first power of u_n appears at each side of the equation then if $u_n(k_1, \omega_1)$ is a solution and $u_n(k_2, \omega_2)$ is another solution, any linear combination of $u_n(k_1, \omega_1)$ and $u_n(k_2, \omega_2)$ is also a solution. The general solution can be written in the form $A_1 u_n(k_1, \omega_1) + A_2 u_n(k_2, \omega_2)$. No matter what the temperature or how many phonons are present in the crystal there can never be any interaction between them in the harmonic approximation.

This same statement applies also to a three-dimensional crystal. Phonons added to a face A of a crystal (see Fig. 2.10) in which the restoring forces are harmonic will cross to a face B without being scattered except by the crystal surfaces or by defects. If phonons are added at a constant rate at one side of the crystal and removed at the other side, then once equilibrium has been reached (the phonons take a finite time to cross AB) the rate at which phonons leave face A will be equal to the rate at which they reach face B. The width AB of the crystal will have no effect on the energy flow and there will be no need for a non-zero temperature difference between the faces. The thermal conductivity will be infinite.

We know that the thermal conductivity of real crystals is finite. There must be some anharmonic terms in the crystal potential to produce this effect. Let us return to our one-dimensional model. The reason why we can get a solution (valid for all values of n and for all t) of the form $\exp\{i(kna - \omega t)\}$ to eq. (2.3) is because the factors $\exp\{i(kna - \omega t)\}$ cancel through eq. (2.4).

Suppose that eq. (2.3) is replaced by

$$m \frac{\partial^2 u_n}{\partial t^2} = f(u_{n+1} + u_{n-1} - 2u_n) + h(u_{n+1} + u_{n-1} - 2u_n)^2. \tag{2.6}$$

With a simple 'ball-and-spring' model there can be no terms on the right-hand side of the equation which depend on an even power of the displacements. However, such terms do exist in real three-dimensional

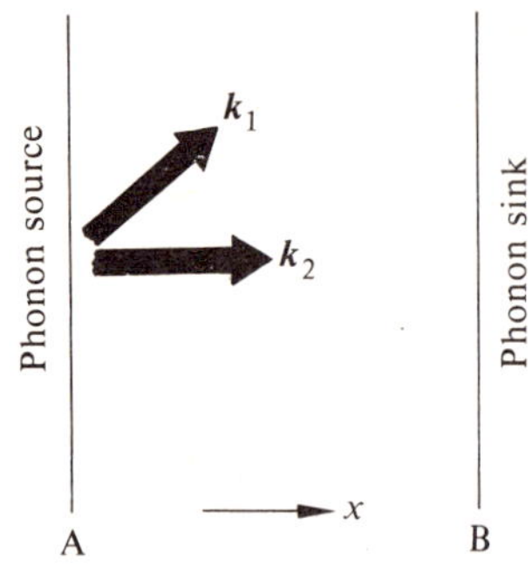

Fig. 2.10. In the harmonic approximation, phonons with wave vector k_1 cannot interact with phonons of wave vector k_2.

crystals and we can get an idea of what happens to the wave motion in these crystals by using eq. (2.6) without worrying about how the terms in h actually arise.

$A \exp\{i(kna - \omega t)\}$ is no longer a solution of the equation of motion. Substitution of this expression into eq. (2.6) gives terms on the right-hand side which depend on $\exp\{i(2kna - 2\omega t)\}$ and there are no terms on the left-hand side to balance them. Terms of the form $\exp\{i(2kna - 2\omega t)\}$ must appear in the true solution. Let us try

$$A[\exp\{i(kna - \omega t)\} + \beta \exp\{i(2kna - 2\omega t)\}] \text{ where } \beta \text{ is small.}$$

Then the addition to the right-hand side of the equation of motion is

$$Af\beta \exp\{i(2kna - 2\omega t)\}\{\exp(i2ka) + \exp(-i2ka) - 2\}$$
$$+ A^2 h \exp\{i(2kna - 2\omega t)\}\{\exp(ika + \exp(-ika) - 2\}^2$$
$$+ \text{smaller terms (of the order of } h\beta).$$

The addition to the left-hand side of the equation is

$$-m(2\omega)^2 \beta A \exp\{i(2kna - 2\omega t)\}.$$

Therefore, for a solution,

$$-4m\beta\omega^2 = -4f\beta \sin^2(ka) + Ah\{2\cos(ka) - 2\}^2.$$

Substituting for $m\omega^2$

$$\beta[8f\{\cos(ka) - 1\} + 4f\sin^2(ka)] = Ah\{2\cos(ka) - 2\}^2$$
$$-4\beta f\{\cos(ka) - 1\}^2 = Ah\, 4\{\cos(ka) - 1\}^2$$
$$\beta = -Ah/f.$$

We still do not know the *exact* solution of eq. (2.6) because higher order terms have been ignored.

Now what happens if we wish to consider two modes (k_1, ω_1) and (k_2, ω_2) vibrating together? The expression $A_1 \exp\{i(k_1 na - \omega t)\} + A_2 \exp\{i(k_2 na - \omega t)\}$ is no longer a solution. By analogy with the argument above, terms of the form $\exp\{i(2k_1 na - 2\omega_2 t)\}$, $\exp\{i(2k_2 na - 2\omega_2 t)\}$, and $\exp[i\{(k_1 + k_2)na - (\omega_1 + \omega_2)t\}]$ must appear in the true solution.

The probability that any transition will take place is proportional to the square of the overlap integral between the initial and final states. In the harmonic approximation the probability that a phonon (k_1, ω_1) will combine with (k_2, ω_2) to give a phonon (k_3, ω_3) depends on the square of

$$\sum_n [A_1 \exp\{-i(k_1 na - \omega_1 t)\} + A_2 \exp\{-i(k_2 na - \omega_2 t)\}]$$
$$A_3 \exp\{-i(k_3 na - \omega_3 t)\}.$$

If k_3 is not equal to either $k_1 (+2\pi r/a)$ or $k_2 (+2\pi r/a)$ (which means a transition of a state into itself) then this summation when taken over a large number of atoms is zero.

When anharmonic terms are considered

$$\langle i \mid f \rangle \begin{array}{c}\text{will have terms}\\ \text{of the form}\end{array} \sum_n \exp\{-i(k_1 na - \omega_1 t)\} \exp\{-i(k_2 na - \omega_2 t)\}$$
$$\exp\{i(k_3 na - \omega_3 t)\}$$

which will be non-zero only if

$$k_3 = k_1 + k_2 + 2\pi r/a \quad (r = \text{any integer}).$$

It will be non-zero for all t if

$$\omega_3 = \omega_1 + \omega_2 .$$

This second equality is just the law of conservation of energy, for the energy associated with a phonon of frequency ω is given by $\hbar\omega$. Such phonon interactions cannot take place in our one-dimensional lattice because the relationship between ω and k is not linear. It is not possible to satisfy the two conditions simultaneously.

However, in three dimensions we have the conditions

$$\boldsymbol{k}_3 = \boldsymbol{k}_1 + \boldsymbol{k}_2 + \boldsymbol{G} \tag{2.7}$$

where $\boldsymbol{G}$ is any reciprocal lattice vector (including the zero vector)

$$\hbar\omega_3 = \hbar\omega_1 + \hbar\omega_2$$

and it is possible to find values of the ks and ωs which satisfy both equations. Anharmonic effects in crystals provide a mechanism for phonon interaction. Do all such processes help to provide a thermal resistivity?

To find which processes have important effects on the thermal conductivity we must consider eq. (2.7) in more detail. The phonons k_1 and k_2 are added to the face A of the crystal of Fig. 2.10. Without loss of generality we can take k_1 and k_2 to be within the first Brillouin zone. k_1 and k_2 are such that their x-components k_{1x} and k_{2x} are positive, i.e., the phonons are moving towards face B. Now the possible combinations of k_1 and k_2 are of two types.

1. The value of G necessary to yield a k_3 in the first Brillouin zone is the zero vector.

$$k_{1x} + k_{2x} = k_{3x}$$

and k_3 is in the first zone so that a positive k_{3x} means that the phonon (k_3) is also moving towards B.

Processes in which $G = 0$ are known as N-processes (normal processes) and are the only ones to take place when k_1 and k_2 are small. See Fig. 2.11(a). If these were the only processes available, the thermal conductivity would be infinite for the following reason. A set of phonons leave face A of the crystal and the total 'momentum'

$$J = \sum_k \hbar k n_k$$

(where n_K = the number of phonons with wave vector k) is not equal to zero. There is a finite J in the positive x-direction. N-processes will leave J unchanged but they will rearrange the phonon distribution to give the most probable distribution compatible with the particular value of J. Some phonons may be produced which are moving back towards face A. The most probable distribution will be reached after a few phonon collisions. Then, because J is non-zero there will be more phonons moving in the positive x-direction than in the negative x-direction and there will be a net energy transfer in the positive x-direction which does not need a finite temperature gradient to maintain it.

It is worth noting that, for an elastic continuum – if such a thing were to exist – we would have

$$k_3 = k_1 + k_2$$

and N-processes would be the only ones possible. Even with anharmonic terms one would still have an infinite thermal conductivity.

2. When k_1 and k_2 get larger and approach the edge of the zone then $k_1 + k_2$ may be outside the zone boundary. In order to get a k_3 within the first zone, G will have to be a reciprocal lattice vector other than zero. Such processes are known as U-processes (from the German word Umklapp) and for a two-dimensional case are illustrated in Fig. 2.11(b). These only happen for a discrete lattice where we have the situation that a large positive value of $k_x{}'$, the x-component of a wave vector k' outside the first zone, may actually correspond to a wave moving in the negative x-direction. It is the U-processes that cause back-scattering of the phonons and lead to the observed finite conductivity – at any rate in defect-free crystals at normal temperatures.

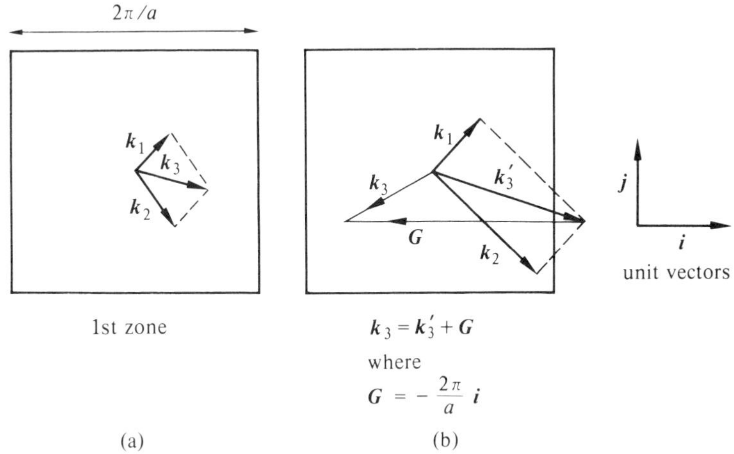

Fig. 2.11. (a) N-process. (b) U-process. In both cases the zone illustrated is that of a simple square lattice of spacing *a*.

At very low temperatures there are very few phonons in the modes with wave vectors near the zone boundary for these modes have high frequency and hence phonons of high energy. U-processes are therefore very improbable and conductivity takes place by 'radiation' of phonons. Any thermal resistance must be due to scattering by defects (including isotopes) or by the crystal boundaries. At these low temperatures the N-processes can have an effect on the thermal resistance – not because they themselves scatter the phonons back but because they can alter the way defect and boundary scattering occur.

It should be mentioned that the probability of the combination $k_1 + k_2 \rightarrow k_3$ is exactly the same as the probability of the dissociation. $k_3 \rightarrow k_1 + k_2$ and that other types of interaction involving a greater number of phonons can also take place.

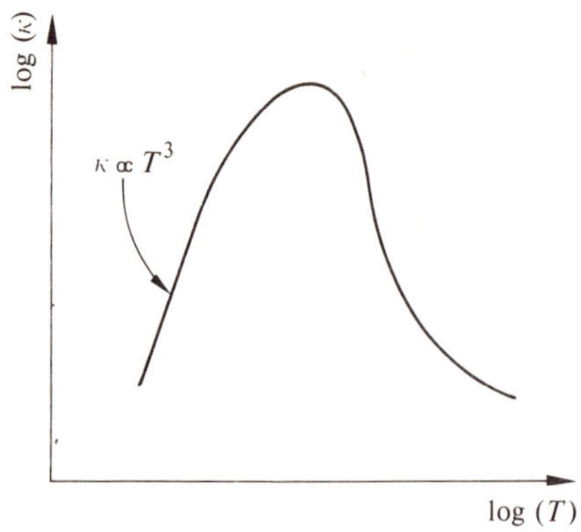

Fig. 2.12. A sketch of the dependence of the thermal conductivity of an insulator upon the absoulte temperature. At low temperatures the size effect becomes important and κ is proportional to T^3.

Figure 2.12 shows the observed variation with temperature of the thermal conductivity of a pure insulator. The definition of thermal conductivity κ is

$$\frac{\partial Q}{\partial t} = A\kappa \frac{\partial T}{\partial x}$$

where $\partial Q/\partial t$ is the rate of transfer of heat energy across an area A when there is a temperature gradient of $\partial T/\partial x$. One would expect (see also p. 96) that the conductivity would be directly proportional to the velocity of the phonon, to the mean free path of the phonon, and also to the specific heat which is a measure of the way the density of phonons changes with temperature. As the temperature is lowered from, say, room temperature, the conductivity increases because the mean free path of the phonons increases. It does not increase indefinitely. Eventually the mean free path is equal to the specimen diameter and can increase no further. κ then behaves in much the same way as does the specific heat and shows a T^3 dependence.

Crystal momentum

For a single particle with linear momentum p the associated wavelength is given by $\lambda = h/p$. Then the wave vector $k = 2\pi/\lambda = p/\hbar$. Here we have a direct relationship between the momentum and the wave vector.

We can define the 'crystal momentum' associated with a phonon of wave vector k as $\hbar k$. Of course 'ordinary momentum' is not carried through the crystal when a vibrational wave passes through it. The atoms all vibrate about their equilibrium positions. Does crystal momentum behave at all like ordinary momentum? Does it, for instance, obey the law of conservation of momentum?

When two phonons collide the conservation law is

$$k_3 = k_1 + k_2 + G$$

which can be taken as a law of conservation of momentum if G is zero, i.e., for N-processes.

The inelastic scattering of slow neutrons (which do have ordinary momentum) gives further evidence for the association of some kind of momentum with the crystal wave vectors. When an incident neutron interacts with a crystal to produce a phonon of wave vector $\boldsymbol{k}$ the conservation law is

$$\underbrace{\hbar K}_{\text{momentum of neutron before collision}} = \underbrace{\hbar K'}_{\text{momentum of neutron after collision}} + \underbrace{\hbar k}_{\text{crystal momentum}} + \hbar G.$$

Examples

1) 1964 Paper 1B

Estimate the Debye θ of gold from the following information. The atomic weight of gold is 197, its density is 19.3, and the velocity of sound in gold is 2100 metres per second. The atomic weight of copper, which has the same crystal structure as gold, is 63.5, its density is 8.9, and the velocity of sound in copper is 3800 metres per second. The Debye θ of copper is 348 K.

Answer

The Debye temperature θ_D is related to ω_D, the Debye cut-off frequency, by $\hbar\omega_D = k_B\theta_D = 6\pi^2(N/V)v^3$ where N/V is the number of atoms per unit volume and v is the velocity of elastic waves in the solid. Therefore

$$\frac{\theta_D(\text{gold})}{\theta_D(\text{copper})} = \frac{\omega_D\,(\text{gold})}{\omega_D\,(\text{copper})} = \left(\frac{N/V\,(\text{gold})}{N/V\,(\text{copper})}\right)^{1/3} \times \frac{\text{velocity of sound in gold}}{\text{velocity of sound in copper}}$$

$$\frac{\text{The number of atoms of gold per unit volume}}{\text{The number of atoms of copper per unit volume}} = \frac{19.3/197}{8.9/63.5}$$

$$\frac{\theta_D\,(\text{gold})}{\theta_D\,(\text{copper})} = \left(\frac{19.3/197}{8.9/63.5}\right)^{1/3} \times \frac{2100}{3800}$$

which gives $\theta_D(\text{gold}) \sim 170$ K.

2) 1970 Paper IVA

At low temperatures the Debye temperature of NaCl and KCl, which have the same crystal structure, are 310 K and 230 K respectively. The lattice specific heat of KCl at 5 K is 3.8 10^{-2} Joule $mole^{-1}$ $degree^{-1}$. Estimate the lattice specific heat of NaCl at 5 K and of KCl at 2 K.

Answer

At low temperatures C_v is proportional to (T^3/θ_D) times the number of vibrational modes per mole. A mole of KCl contains the same number of atoms as does a mole of NaCl.

$$\frac{C_v\ (\text{NaCl})\ \text{at 5 K}}{C_v\ (\text{KCl})\ \text{at 5 K}} = \frac{\theta_D{}^3\ (\text{KCl})}{\theta_D{}^3\ (\text{NaCl})} = \frac{(230)^3}{(310)^3}$$

$$C_v\ (\text{NaCl})\ \text{at 5 K} = \frac{230}{310} \times 3.8 \times 10^{-2}\ \text{J mole}^{-1}\ \text{degree}^{-1}.$$

$$\frac{C_v\ (\text{KCl})\ \text{at 2 K}}{C_v\ (\text{KCl})\ \text{at 5 K}} = \frac{2^3}{5^3}$$

Therefore,

$$C_v\ (\text{KCl})\ \text{at 2 K} = \frac{8}{125} \times 3.8 \times 10^{-2}\ \text{J mole}^{-1}\ \text{degree}^{-1}.$$

Questions

1) Bristol, 1974

Calculate the zero-point energy for a three-dimensional Debye solid. Estimate the amplitude of the zero-point vibrations for a crystal with atoms of atomic mass 40 and Debye temperature 100 K.

[1 a.m.u. = 1.66 10^{-27} kg.]

2) Manchester, 1971

Define the Debye temperature of a crystalline solid and comment on its significance in the temperature variation of specific heat.

Potassium has the body-centred-cubic structure with the unit cell side 5.33 Å. Calculate the Debye temperature on the assumption that transverse and longitudinal elastic waves in potassium have the common velocity 1.5 10^3 m s^{-1}.

3) Bristol, 1973

The low temperature specific heat of copper varies as $AT + BT^3$, with $B = 4.6\ 10^{-5}$ joule $mole^{-1}$ deg^{-4}. What Debye temperature θ does this correspond to? If the velocities of longitudinal waves and shear waves in copper are 3.75 and 2.17 10^5 cm s^{-1} respectively, what value of θ would the Debye theory predict? What would be the cut-off frequency ν_{max} on the Debye theory, and the corresponding minimum wavelengths λ_{min} of longitudinal and shear waves?

Show that if the Debye theory is modified by making ν_{max} different for longitudinal modes and shear modes, the values of λ_{min} become equal. At low temperatures, the specific heat will still vary as T^3: what will now be the effective value of θ?
[You need not worry about the AT term which is due to the conduction electrons – see Chapter 4.]

4) Manchester, 1973
Discuss the lattice vibrations of a one-dimensional chain of N identical atoms, each of which has a mass m.

The interatomic potentials between two atoms at a separation r in such a chain is

$$V(r) = \frac{-A}{r^6} + \frac{A\,a^6}{2r^{12}},$$

where a is the equilibrium interatomic spacing. Calculate the velocity of sound in the chain and the maximum frequency of lattice vibrations of the chain.
(Consider only the interaction between nearest neighbours in the chain.)

5) Manchester, 1972
Describe the assumptions of the Debye model of the lattice heat capacity C_L of a solid and indicate how successful it is in accounting for experimental measurements of C_L.

Using the assumptions of the Debye model obtain an expression for the lattice vibrational energy per unit interval of angular frequency in a solid in thermal equilibrium at temperature T. Show that this quantity has a maximum at angular frequency ω_M where

$$\exp\left(\frac{\hbar\omega_M}{kT}\right)\left(1 - \frac{1}{3}\frac{\hbar\omega_M}{kT}\right) = 1.$$

Hence show that the wavelength of the phonons at this maximum is of the order $a\,\theta/T$, where a is the interatomic distance and θ is the Debye temperature.

3. Ionic Crystals

Cohesive energy of ionic crystals

The ionic bond was mentioned in chapter 1. In some crystals, for example the alkali halides, the *main* bonding mechanism is the ionic bond. If we ignore other types of bonds and assume that there is complete transfer of the outer electrons (from the metal atoms to the halogen atoms in the alkali halides) then we can make some fairly simple calculations of the binding energy. The ions, then, are considered to have noble-gas electronic configurations and such arrangements of the electron charge are spherically symmetric. A spherically symmetric charge distribution can be treated as a point charge at the centre of the distribution. Isolated ions would therefore produce an electrostatic potential outside the ion like that of a point charge. We must remember that the ions in a crystal are not isolated and the electron cloud will be affected by the nearest neighbour ions. X-ray pictures show this distortion but effects due to polarization of the ions can be ignored for the purposes of the simple calculations which follow.

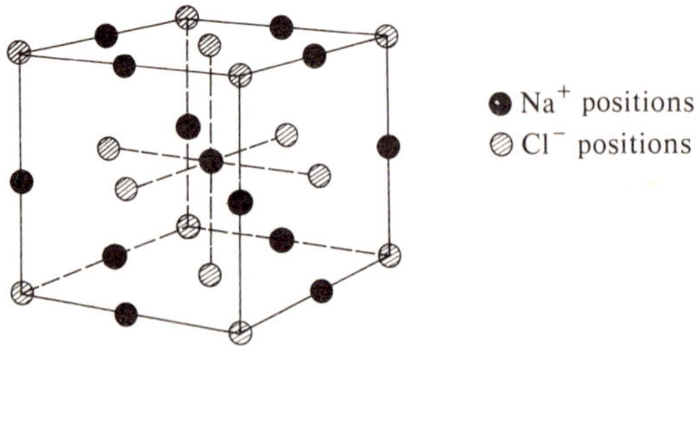

Fig. 3.1. Equilibrium positions of ions in sodium chloride crystal.

Let us consider the binding energy of sodium chloride. We need to know the energy of the reference ion (see Fig. 3.1) in the crystal as compared with its energy an infinite distance away from all the other ions. The reference ion has a positive charge. There are six nearest neighbours at sites like $a\ 0\ 0$ where a is the interionic spacing. These have negative charges. There are twelve neighbours at sites like $a\ a\ 0$ with positive charges and eight at sites like $a\ a\ a$ with negative charges, etc. If we write the charge on the ion as q (which is generally the electronic charge x the number of units of charge on the ion) we find that the electrostatic energy must be expressed in terms of a series:

$$E_{\mathrm{el}} = \frac{-q^2}{4\pi\epsilon_0 a}\left(6 - \frac{12}{\sqrt{2}} + \frac{8}{\sqrt{3}} - \ldots\right).$$

The term in brackets is known as the Madelung constant (α) and depends only on the *type* of crystal involved. We know that α must be positive otherwise the potential energy of the ion in the crystal would be greater than its energy at infinity and the Coulomb forces would push the crystal apart rather than pull it together. The actual values of α are not very easy to calculate by summing such a series, for it converges only very slowly – at greater distances there are a greater number of ions. Methods of calculating α often make use of neutral shells of atoms (or fractions of atoms) and involve the sums of series which converge more rapidly. Values of α for three simple crystal types are shown in Table 3.1.

TABLE 3.1

Crystal structure	Madelung constant
Rocksalt structure	1.748
Caesium chloride structure	1.763
Zinc blende structure	1.638

The potential energy of the reference ion due to the Coulomb forces between it and all the other ions is therefore

$$E_{\mathrm{el}} = -q^2\alpha/a\ 4\pi\epsilon_0.$$

The problem now is to find an expression for the repulsive potential. The repulsive forces are short-range forces which keep the crystal from collapsing indefinitely. When the distance r between the ions is much greater than $r_+ + r_-$, where r_+ is the 'radius' of the cation and r_- the 'radius' of the anion, then the repulsive potential is *very* small. (More about the 'radii' of ions will be found on p. 72). As the ions are moved closer together this potential remains small until they 'touch' when it begins to increase rapidly. The ions behave like 'fairly hard' spheres – they can be compressed only slightly.

We need an expression for the repulsive potential which is reasonably accurate – at least in the region around $r = a$. The repulsion is not a classical phenomenon; it is due to quantum-mechanical effects. An exact expression, if we were able to derive one, would be very complicated and would vary slightly for every possible pair of ions. A simple expression due to Born is

$$E_{\text{rep}} = \beta/r^n (= \beta/a^n \text{ at the equilibrium spacing})$$

where n is large to give the necessary short-range behaviour.

When the crystal is in equilibrium ($r = a$) the potential energy must be a minimum. The two terms – electrostatic and repulsive – and the total potential energy E are shown in Fig. 3.2.

$$E = \frac{-\alpha q^2}{4\pi\epsilon_0 r} + \frac{\beta}{r^n} \tag{3.1}$$

$$\frac{\partial E}{\partial r} = \frac{+\alpha q^2}{4\pi\epsilon_0 r^2} - \frac{n\beta}{r^{n+1}} \tag{3.2}$$

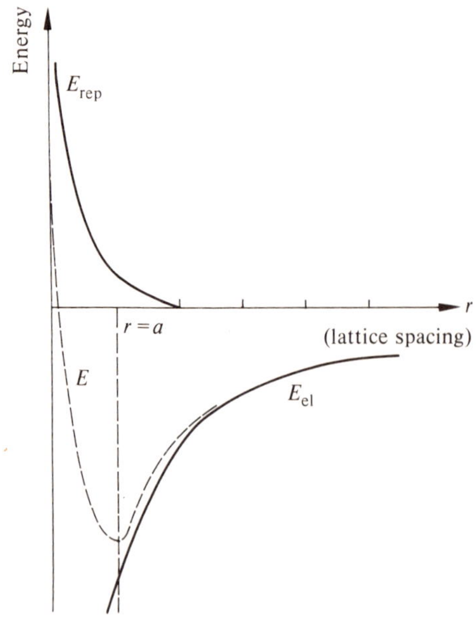

Fig. 3.2. Potential energy of an ion in a crystal.

and this must be zero when $r = a$. So, we have

$$\alpha q^2/a^2 4\pi\epsilon_0 = n\beta/a^{n+1} \quad \text{or} \quad \beta = a^{n-1}\alpha q^2/n4\pi\epsilon_0. \tag{3.3}$$

The repulsive coefficient, β, can be found if the Born exponent n is known. n can be deduced from compressibility data, as will be explained later. It is found that, for all crystals, n is approximately equal to 9, but a better approximation is to make use of a table produced by Pauling.[1]

TABLE 3.2

Ion type	Born exponent
Helium	5
Neon	7
Argon	9
Krypton	10
Xenon	12

This table, which is reproduced in Table 3.2, gives a relationship between the ion type and n. As an example of the way to use this table let us consider lithium fluoride. The electronic configuration of Li^+ is that of the helium atom and that of F^- is of the neon atom: we can get an approximate value of the Born exponent in the crystal LiF by taking the average value of five and seven, that is six.

The potential energy (relative to separation into *ions*) of a crystal of N positive and N negative ions is therefore

$$E_N = N\left\{\frac{-\alpha q^2}{a4\pi\epsilon_0} + \frac{n\beta}{a^{n+1}}\right\}$$

$$= N\left\{\frac{-\alpha q^2}{a4\pi\epsilon_0} + \frac{\alpha q^2}{na}\right\}$$

$$= \frac{-N\alpha q^2}{a4\pi\epsilon_0}\left\{1 - \frac{1}{n}\right\}. \tag{3.4}$$

Notice that the term in brackets is multiplied by N and not by $2N$ because this would mean that the potential energy between each pair of ions was counted twice. The first term on the right-hand side is E_{el} and the second term is E_{rep}. E_{rep} is only $1/n \times E_{el}$ and n is large (~9) so the main contribution to the cohesive *energy* is from electrostatic terms. The electrostatic and repulsive *forces* are *equal* at the equilibrium spacing.

The equilibrium lattice spacing, a, can be found from crystallographic measurements using X-rays. n can be deduced in the following way from data on the compressibility of the crystal. The volume of a crystal containing N ion pairs must be proportional to r^3 and to N. It can be written as CNr^3 where C depends only on the type of lattice. The compressibility is defined by

$$k = -\frac{1}{v}\left(\frac{\partial V}{\partial p}\right)_T$$

The change in energy when a crystal changes its volume by ∂V is given by $\partial E = -p\,\partial V$.

$$p = -\left(\frac{\partial E}{\partial V}\right)_T,$$

therefore

$$\left(\frac{\partial p}{\partial V}\right)_T = -\left(\frac{\partial^2 E}{\partial V^2}\right)_T$$

$$\frac{1}{kV} = \left(\frac{\partial^2 E}{\partial V^2}\right)_T. \qquad (3.5)$$

Using eq. (3.1) multiplied by N

$$\left(\frac{\partial E_N}{\partial V}\right)_T = \frac{\mathrm{d}r}{\mathrm{d}V}\left(\frac{\partial E}{\partial r}\right)_T = \frac{1}{3CNr^2}N\left\{\frac{\alpha q^2}{r^2 4\pi\epsilon_0} - \frac{n\beta}{r^{n+1}}\right\}.$$

$$\left(\frac{\partial^2 E_N}{\partial V^2}\right)_T = \frac{1}{(3CNr^2)^2}N\left\{\frac{-2\alpha q^2}{4\pi\epsilon_0 r^2} + \frac{n(n+1)\beta}{r^{n+2}}\right\}.$$

Using eqs. (3.3) and (3.5) and putting $r = a$, the equilibrium spacing, we get,

$$n = 1 + 4\pi\epsilon_0 9Ca^4/k\alpha q^2.$$

In this formula, k is the compressibility corresponding to the equilibrium spacing, i.e., the compressibility at absolute zero. Experimental measurements at low temperatures are extrapolated to $T = 0$ K.

The Born theory which has been outlined and which makes use of an electrostatic attractive potential and a simple repulsive potential depending on r^{-n} can be improved by

1. Including an energy term due to the Van der Waals forces which exist between the ions, and
2. Taking into account the contribution from the zero-point energy of the crystal. Also:
3. Quantum mechanical calculations indicate that the repulsive potential will be better approximated by a term such as

$$E_{rep} = \gamma \exp(-r/\rho).$$

The cohesive energy mentioned in chapter 1 is taken to be positive and is therefore $-E$ per ion pair. We have seen how this can be calculated if it is assumed that the bonding is ionic with no covalent contribution. The cohesive energy can also be measured experimentally by a Born–Haber cycle, which will not be discussed here but which can be found in the standard texts of solid-state physics and also in books on chemical thermodynamics. No compound is purely ionic and it might be thought that a comparison of experimental and theoretical results would yield an expression for the amount of covalent character in the crystal bonds. Unfortunately there is no simple relationship between the amount of covalent character and this difference because the ionic model is *self-compensating*. If there is some covalent bonding then the charge transfer is not complete and by working with formal charges as in the preceding calculation we arrive at an ionic contribution to the binding energy which is greater than the correct one. The difference between experimental and theoretical values of the cohesive energy is therefore less than (and maybe much less than) the contribution of covalent bonding to the energy.

The cohesive energy which we have calculated is the energy relative to *ions* at infinite separation but there is another way of defining the cohesive energy – namely, relative to *neutral atoms* at infinite separation. The difference between these energies depends on the ionization energy and the electron affinity of the atoms concerned. Let us again consider the sodium chloride crystal. In order to separate the crystal into ions at infinite separation we need an energy E (see eq. (3.1)) per ion pair. Then, in order to change these ions to neutral atoms we must remove an electron from a chloride ion and add it to a sodium ion. The amount of energy needed to remove an electron from a Cl^- ion is 3.61 eV – the electron affinity of chlorine. The energy gained when an electron is added to a Na^+ ion is 5.14 eV – the ionization energy of sodium. Note that 5.14 eV is greater than 3.61 eV – in other words the energy of the two atoms at infinite separation is less than the energy of the two ions at infinite separation. This result is not restricted to sodium chloride – chlorine has the highest electron affinity of the elements and caesium, which is the element with the lowest ionization energy, still needs 3.89 eV to enable it to give up an electron. Even in the most favourable circumstances energy is needed to produce an ion pair at large distances apart. Note, also, that if the distance r between a pair of ions decreases from infinity, then there comes a distance r_0 below which the ionic state would have the lower energy. A rough value of r_0 is found by equating the drop in potential energy due to Coulomb forces, $q^2/4\pi\epsilon_0 r_0$, with the energy needed to produce an ion pair. Using our example of sodium chloride:

$$\frac{q^2}{4\pi\epsilon_0 r_0} = 5.14 - 3.61 \text{ eV} \quad \text{or} \quad r_0 = 0.94 \text{ nm } (0.94\ 10^{-9} \text{ m}).$$

(The nearest-neighbour spacing in the *crystal* is 0.28 nm. We have assumed that the electron affinity and the ionization energy are not influenced by the other atom nearby and repulsive effects have been ignored.)

The cohesive energy (ions) is greater than the cohesive energy (atoms) and this in turn must be greater than zero otherwise the crystal would not be stable. For sodium chloride:

$$E(\text{ions}) - E(\text{atoms})$$
$$= \text{ionization energy of Na} - \text{electron affinity of Cl.}$$

Ionic radii and their influence on crystal structure

Ions do not behave like hard spheres. The 'radius' of an ion has no absolute significance but simply depends on the definition of the word and the way in which the measurement or calculation is carried out. Quantum mechanical calculations show that the electron density around an isolated ion decreases exponentially with the distance from the ion (at large distances) but it only reaches zero when the distance is infinite. Is there any point in talking about the 'radius' of an ion in a crystal? We can say that the equilibrium interionic distance in a crystal is equal to the sum of the radii of the anion and of the cation. However, this definition will be of very little use unless it can be shown that the radius of a certain ion does not depend strongly on the particular crystal that it happens to be in. Experimentally it is found that, to a reasonable approximation, ionic radii are additive quantities. For example, if we calculate the difference between the radii of the potassium and sodium ions ($r_{K^+} - r_{Na^+}$) from the interionic distances in the crystals of the halides of these metals we find that ($r_{K^+} - r_{Na^+}$) is constant within about 15 per cent. The interionic spacings can, of course, be measured by X-ray diffraction methods.

Comparisons such as these will yield a table of ionic radii provided that the radius of one ion is known. Goldschmidt in 1927 produced a table of ionic radii based on the assumption that the radius of F^- was 1.33 Å (0.133 nm). This value for the radius of F^- was calculated from an equation relating polarizability and ionic 'size'.

Pauling produced another table. He considered crystals composed of isoelectronic ions such as NaF, KCl, RbBr, and CsI. (Na^+ and F^- are isoelectronic because they both have the same number of electrons – ten in this case, i.e., the electronic configuration of the neon atom.) Pauling said that the radius of an ion should be inversely proportional to the effective nuclear charge as seen by the outermost electrons. $Z_{eff} = Z - S$ where S is a screening factor determined partly by theoretical calculations and partly by experimental measurements. S depends on the electronic configuration but not on the nuclear charge and therefore S is the same for the isoelectronic ions. Also for such ions the constant of proportionality is the same.

$$\text{Radius} = \frac{C}{Z_{eff}} = \frac{C}{Z - S}.$$

Taking the case of sodium fluoride, we can find $r_{Na^+} + r_{F^-}$ from X-ray measurements and if S has been calculated then, knowing $Z_{Na} = 11$ and $Z_F = 9$, we can find r_{Na^+}/r_{F^-} and hence the two radii can be found. In a similar way, actual values (and not just differences) of the radii for K^+, Cl^-, etc., can be calculated. Here we have the basis of a table of ionic radii.

The tables produced by different workers do not agree exactly – which is hardly surprising – but the disagreement is not too bad (e.g., Pauling's value for F^- was 1.36 Å).

Data found in text books are self-consistent values of the radii which when added together give reasonably correct values for interionic spacing in the majority of crystals at room temperature. The values are usually quoted for six-coordination. The rocksalt structure is such that each ion has six nearest neighbours of opposite sign. This crystal type is said to have six-coordination. The coordination number for the caesium chloride structure is eight, and for the zinc blende structure is four.

It can easily be shown that the radius of an ion would be expected to depend on the coordination number. From eq. (3.3)

$$a^{n-1} = \frac{\beta n 4\pi\epsilon_0}{\alpha q^2} \quad \text{or } a = \text{equilibrium spacing} = \left(\frac{4\pi\epsilon_0 n\beta}{\alpha q^2}\right)^{1/(n-1)}$$

If n is given the average value of nine and β is taken to be proportional to the number of nearest neighbours – a fairly sensible assumption – then

$$\frac{a\,(\text{CsCl})}{a\,(\text{NaCl})} = \left(\frac{8}{6}\,\frac{1.748}{1.763}\right)^{1/8} = 1.03$$

$$\frac{a\,(\text{Zns})}{a\,(\text{NaCl})} = \left(\frac{4}{6}\,\frac{1.748}{1.638}\right)^{1/8} = 0.958$$

Thus, the equilibrium spacing and hence the effective radii of the ions change as the crystal type changes. As the coordination number changes from eight to four the radii decrease by about seven per cent.

Isn't there some way of measuring the radii directly? Studies of the electron-density maps obtained from X-ray analysis could be used. Unfortunately the positions of minimum electron density often give values for the ionic radii which vary drastically from those in the tables. The tabulated values have no absolute meaning but are useful because *on average* they give good results in calculations of properties which depend on the size of the ions.

Radius-ratio rules. Crystals of the type MX where M is a cation and X an anion exist in different crystal types such as rocksalt, caesium chloride, and zinc blende structures. Can we explain why a compound should choose to crystallize in one particular form rather than in another? Obviously the form with lowest energy will be chosen. In order to arrive at an explanation for the existence of the preferred structure we assume that

1. the ions behave as hard spheres, the radii fixed and independent of coordination number, and

2. the bonding is purely ionic and the lattice energy can be expressed in the form

$$E_N = \frac{-N\alpha q^2}{a 4\pi\epsilon_0}\left\{1 - \frac{1}{n}\right\}.$$

Now, the crystal will be as tightly bound as possible if α is large and a is small. The largest value of α (for the three structures mentioned above) is that for the caesium chloride structure, i.e., $\alpha = 1.763$.

In practically all cases, anions are larger than cations. So let us say that $r_M < r_X$ and that we will consider r_X to be fixed. As r_M/r_X decreases from unity, $a = r_M + r_X$ decreases until a critical value is reached when the anions touch. (See Fig. 3.3(a).) This happens when

$$r_M + r_X = \text{half body-diagonal of cube} = \frac{\sqrt{3}}{2} \times \text{side of cube}$$

$$= \frac{\sqrt{3}}{2}\, 2r_X.$$

The interionic spacing $a = r_M + r_X$ is now fixed at $\sqrt{(3)}r_X$ and as r_M is made smaller the binding energy will not be affected. The critical ratio,

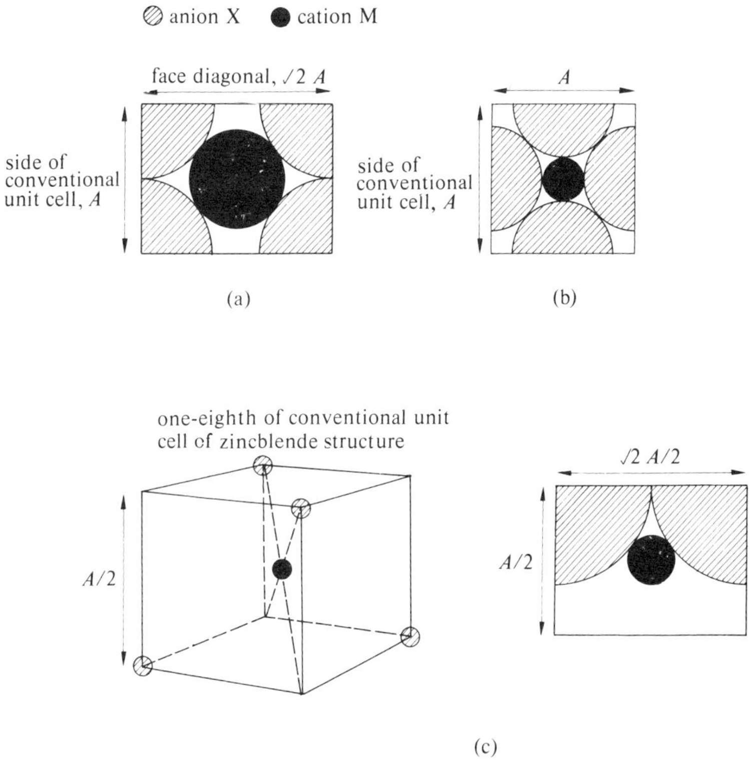

Fig. 3.3. Critical conditions in (a) caesium chloride structure, (b) rocksalt structure, and (c) zinc blende structure.

$$\frac{r_X}{r_M} = \frac{1}{\sqrt{3} - 1} = 1.37.$$

No further increase in the cohesive energy is possible with this structure.

For the sodium chloride structure (Fig. 3.3(b)) the limiting value of a occurs when the anions touch, i.e., when

$$r_M + r_X = \text{half the side of the cube} = \frac{1}{\sqrt{2}} \times 2r_X.$$

An interionic spacing of $\sqrt{(2)}r_X$ is smaller than one of $\sqrt{(3)}r_X$. Therefore, despite the small difference in the Madelung constant, the cohesive energy for a rocksalt-structure crystal can be greater than that for a caesium chloride type crystal if r_X/r_M is large enough. However, the rocksalt structure has a critical value of r_X/r_M above which the cohesive energy cannot increase further.

$$\frac{r_X}{r_M} = \frac{1}{\sqrt{2} - 1} = 2.42.$$

As r_M decreases still further, the zinc blende structure becomes energetically the most favourable despite its small Madelung constant. The limiting value of r_X/r_M (when the anions touch) is not reached until (see Fig. 3.3(c))

$$\sqrt{2}\frac{A}{2} = 2r_X, \quad \frac{A}{2}\sqrt{3} \times \frac{1}{2} = r_M + r_X$$

$$\frac{r_X}{r_M} = \frac{1}{\sqrt{(3/2)} - 1} = 4.45.$$

These ratios must be taken with a pinch of salt. Near the critical ratios the stability of one structure as compared with another may be strongly influenced by the coordination number and its effect on the radius, which we have ignored for this calculation. Polarization effects may also be important enough to alter the above predictions. Some compounds even change from one crystal structure to another as temperature and pressure are varied. However, it is found that *generally* when compounds MX have ions with nearly equal radii, the caesium chloride structure is favoured; when the ionic radii differ by a factor of two or more the zinc blende structure is favoured; in intermediate cases the rocksalt structure occurs. Table 3.3 might be of interest. It shows the actual radius ratio in caesium chloride, sodium chloride, and zinc sulphide – the compounds after which the crystal types are named.

TABLE 3.3

†Radius of cation, r_M		†Radius of anion, r_X		r_X/r_M	Approximate Range expected from calculation
Cs^+	1.69 Å	Cl^-	1.81 Å	1.1	1 – 1.37
Na^+	0.95 Å	Cl^-	1.81 Å	1.9	1.37 – 2.42
Zn^{++}	0.74 Å	S^{--}	1.84 Å	2.5	2.42 – 4.45

†Pauling's values from *The nature of the chemical bond.*[1]

Defects in ionic crystals

So far, I have mentioned only perfect crystals. In fact no crystal can be perfect at finite temperatures. Many defects are possible in ionic crystals. Some of these are:

1. Chemical impurities,
2. Intrinsic point defects such as atoms missing from normally occupied sites (vacancies) or additional atoms in the interstices of the normal lattice (interstitials)
3. Dislocations where whole blocks of a crystal may be moved relative to one another.

One can even consider

4. that the existence of a surface – even if uncontaminated – and
5. the existence of phonons at finite temperatures are defects in a pure, perfect, infinite crystal.

Class (2) will be considered here. Point defects are those defects which do not spread over more than one or two lattice spacings. One kind of point defect is an F-centre – one kind of colour centre. A representation of such a defect is shown for a two-dimensional lattice in Fig. 3.4. A negative ion is replaced by an electron. Electrostatic forces bind the electron to the lattice site. The lattice will be distorted to some extent about the F-centre. The energy levels of the electron are such that electromagnetic radiation in the visible region is absorbed when the electron jumps to a higher energy state. Crystals containing many F-centres are coloured, e.g., sodium chloride turns brown and potassium chloride violet when they are heated in sodium and potassium vapour respectively. Heating in the metal vapour introduces F-centres into the crystals.

Two other common defects are the Frenkel defect and the Schottky defect. These are illustrated schematically in Fig. 3.5. A Frenkel defect occurs when an ion (of either sign) is displaced from its lattice position into an *interstitial* position far enough away from the vacancy for the interaction between the interstitial ion and the vacancy to be ignored.

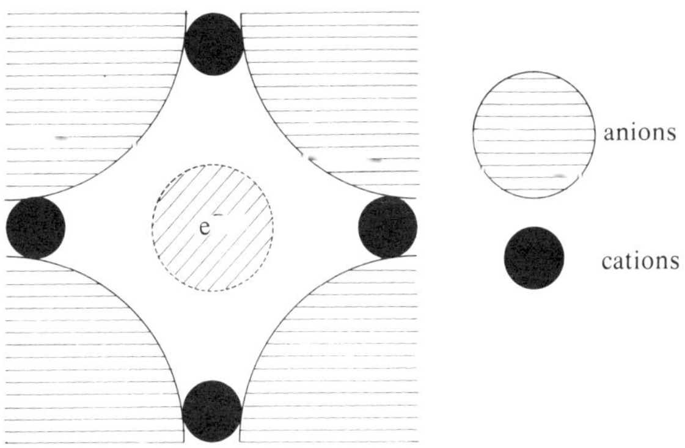

Fig. 3.4. Schematic representation of an F-centre.

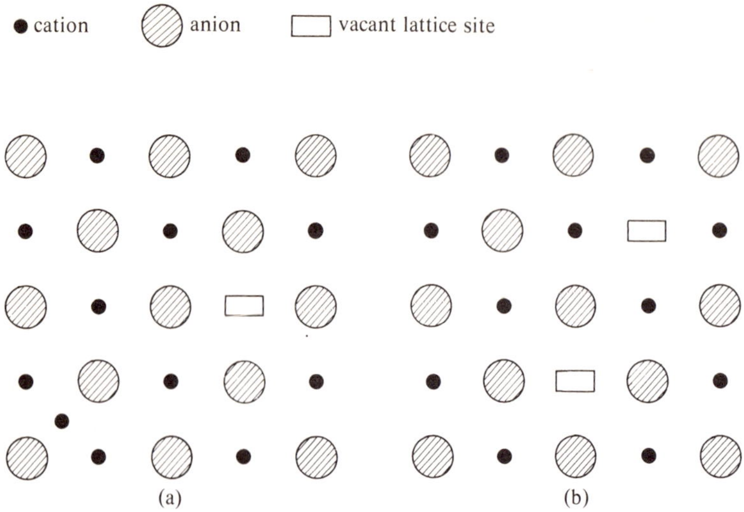

Fig. 3.5. (a) Frenkel defect. (b) Schottky defect. The ions are shown small for clarity.

The neighbouring ions around the interstitial ion will be displaced and deformed slightly but the lattice is back to 'normal' a few lattice spacings away from the defect. For most crystal structures Frenkel defects are more likely to occur in a cation sublattice than in a anion sublattice because the cations are usually smaller than anions and can fit into more gaps in the lattice. Frenkel defects also occur more readily in lattices with low coordination numbers – in such lattices there are more interstitial spaces.

There is a Schottky defect when an ion is actually missing from the lattice. Again, there will be some distortion around the empty lattice site. In ionic crystals of the type MX, positive- and negative-ion vacancies occur in roughly equal numbers to preserve the electric neutrality of the crystal. This happens even if the energy needed to remove a negative ion from the lattice out to the surface, and so create a negative-ion vacancy, is very different from that needed to produce a positive-ion vacancy.

The Frenkel and Schottky defects are in thermodynamic equilibrium with the lattice (in a crystal with a suitable thermal history). One can therefore calculate the number of defects that should be present at a given temperature. The larger dislocations of group (3), should any be present in the crystal, have such large energy barriers to overcome in order to revert to 'normal' that they are usually 'frozen in'. As they are not in thermal equilibrium with the lattice, the following type of calculation cannot be applied to such defects.

The equilibrium concentration of Frenkel defects. Energy is needed to create a Frenkel defect in a perfect lattice. If minimization of energy were the only consideration then these defects would never exist in an equilibrium situation. However, *thermodynamic* equilibrium is obtained, not by minimization of energy, but, for systems held at constant temperature and pressure, by minimization of a function known as the Gibbs energy G.

$$G = E + pV - TS$$

where E is the internal energy of the system, p the pressure, V the volume, T the absolute temperature, and S the entropy. When p is equal to atmospheric pressure, then terms $p\delta V$ associated with a change δG are usually negligible compared with δE and $T\delta S$. In such situations it is reasonable to minimize the Helmholtz energy, F, instead.

$$F = E - TS.$$

An increase in E, which will occur as Frenkel defects are created, will tend to increase F; an increase in the entropy (related to disorder in the crystal) will decrease F. To find when F is a minimum we need to know how S is related to the number n_{Fr} of defects present. Statistical mechanics gives the result that the contribution S_{Fr} to the entropy from the defects is

$$S_{\mathrm{Fr}} = k_{\mathrm{B}} \ln W$$

where W is the number of different ways that the n_{Fr} defects can be arranged in the lattice.

If the number of defects is small so there is no interaction between them, then

$$E_{\mathrm{Fr}} = n_{\mathrm{Fr}} e_{\mathrm{Fr}}.$$

If there are N lattice sites in the sublattice which is subject to defects and N_{i} interstitial sites, then W is calculated in the following way. The number of different ways of arranging $N - n_{\mathrm{Fr}}$ identical ions and n_{Fr} identical vacancies on N sites is

$$\frac{N!}{(N - n_{\mathrm{Fr}})!\,n_{\mathrm{Fr}}!}.$$

The number of different ways of arranging n_{Fr} identical ions and

$N_i - n_{Fr}$ identical vacancies on N_i sites is

$$\frac{N!}{(N_i - n_{Fr})!\, n_{Fr}!}\,.$$

So

$$W = \frac{N!}{(N - n_{Fr})!\, n_{Fr}!} \times \frac{N_i!}{(N_i - n_{Fr})!\, n_{Fr}!}\,.$$

$F = E - TS$ is a minimum when

$$\frac{\partial F}{\partial n} = \frac{\partial E}{\partial n} - \frac{T\partial S}{\partial n} = 0$$

$$= e_{Fr} - k_B T \frac{\partial \ln W}{\partial n_{Fr}}.$$

Using Stirling's approximation (for large N, $\ln N! = N \ln N - N$) and remembering that

$$\frac{\partial}{\partial n} \ln X = \frac{1}{X}\frac{\partial X}{\partial n}$$

$$\begin{aligned}\frac{\partial W}{\partial n_{Fr}} &= \frac{\partial}{\partial n_{Fr}}\{N \ln N - (N - n_{Fr})\ln(N - n_{Fr}) - n_{Fr} \ln n_{Fr} \\ &\qquad + N_i \ln N_i - (N_i - n_{Fr})\ln(N_i - n_{Fr}) - n_{Fr} \ln n_{Fr}\} \\ &= \ln(N - n_{Fr}) + 1 - \ln n_{Fr} - 1 + \ln(N_i - n_{Fr}) + 1 - \ln n_{Fr} - 1 \\ &= \ln\left\{\frac{(N - n_{Fr})(N_i - n_{Fr})}{{n_{Fr}}^2}\right\}.\end{aligned}$$

The equilibrium condition is

$$e_{Fr} = k_B T \ln\left\{\frac{(N - n_{Fr})(N_i - n_{Fr})}{{n_{Fr}}^2}\right\}.$$

$$n_{Fr} = (NN_i)^{1/2} \exp(-e_{Fr}/2k_B T)$$

(n_{Fr} is small in comparison with N and N_i).

This calculation has not taken into account the change in thermal entropy which occurs when defects are produced because of the change in the vibrational modes of the lattice.

The exponential term with $k_B T$ in the denominator is the type of temperature dependence that one expects from systems obeying classical statistics.

The equilibrium concentration of Schottky defects. The analysis is very similar to that for Frenkel defects. Consider a crystal of the type MX. If n_S is the number of Schottky defects present in a crystal with N lattice sites in each sublattice then there will be $N - n_S$ ions of each type.

The number of ways of arranging n_S identical vacancies and $N - n_S$ identical ions on N sites is given by

$$\frac{N!}{n_S!\,(N-n_S)!},$$

so

$$W = \frac{N!}{n_S!\,(N-n_S)!} \times \frac{N!}{n_S!\,(N-n_S)!}$$

and we find that

$$n_S = N \exp\left(-e_S/2k_B T\right)$$

where e_S is the energy needed to produce a pair of vacancies.

The calculations on the equilibrium concentrations of Frenkel and Schottky defects have assumed that only one type of defect is present in the crystal. The assumption is usually justified for e_{Fr} and e_S are usually dissimilar. When e_{Fr} and e_S are nearly equal both types of defect will exist in the same crystal and the calculations will have to be modified.

Ionic conductivity

A perfect, and purely ionic, crystal would not conduct an electric current except perhaps at very large voltage gradients when electrons could be ripped from the ions and pulled through the crystal. Frenkel and Schottky defects provide a mechanism whereby current can flow through a crystal containing them. The interstitial ion of a Frenkel defect can move from one interstitial position to another. Normally it will diffuse through the crystal in a random manner but if an electric field is applied it will have an average motion along the field direction if it is a cation or in the opposite direction should it be an anion. Vacancies also move through the crystal by virtue of the fact that neighbours can move into the vacancies, leaving other vacancies on the neighbouring-ion sites. An electric field will modify the normal random motion of vacancies. A positive-ion vacancy behaves as if it had a negative charge and moves in the opposite direction to the field.

Other defects may be involved in the conduction process. Positive- and negative-ion vacancies are attracted to one another by electrostatic forces and may combine to form a pair of vacancies. A pair of vacancies in contact is not a Schottky defect. Also, if the interstitial ion and the vacancy of a Frenkel defect get sufficiently close together to interact, then they cease to be a Frenkel defect and become a compound defect. Dipositive cations as impurities in alkali halide crystals and their associated vacancies (which must be present in order to preserve electric neutrality of the crystal) may also be involved in conduction.

Conductivity data, taken in conjunction with data from various other types of experiments (of which a few are *very briefly* described here; further information is found in, for example, Greenwood[2]) yield answers to some of the following questions.

In which type of crystals are Frenkel defects more numerous and in which type are Schottky defects predominant? How mobile are the different ions as they move through the crystal either from lattice site to lattice site (possible because of the vacancies present) or by way of interstitial positions? Is ionic conduction due to anions, cations, or both? What are the formation energies e_S and e_{Fr} of the Schottky and Frenkel defects?

The presence of Frenkel defects in a crystal should leave the density unaltered; that of Schottky defects should mean that the crystal is less dense than a perfect crystal although there may be some compensation because of slight collapse of the lattice around the vacancies. The comparison of displacement density with X-ray density can sometimes provide evidence for the identity of defects in a crystal. Unfortunately, although this method is easy to understand in principle, it is difficult in practice because the effect is so small except in grossly defective crystals. Such crystals can be produced by irradiation but the number of defects present depends on the amount of irradiation and not just on the temperature – we do not have a situation of thermodynamic equilibrium.

Theoretical calculations of e_{Fr} and e_S can provide an estimate of the relative numbers of Frenkel and Schottky defects present. For example, the energy of formation e_S of a pair of vacancies is the energy needed to remove a pair of ions from the centre of the crystal to the surface. The expression for the potential energy of a surface ion involves the summation over only one hemisphere of ions – the other hemisphere is outside the crystal. One would therefore expect the cohesive energy for a surface ion to be half that of ions in the interior of the crystal. Roughly, then, e_S should be half the cohesive energy per ion pair. This rough estimate, however, has ignored the polarization effects around the vacancy which tend to reduce the amount of energy needed to form the vacancies. Such polarization effects can be taken into account to give a better value of e_S.

A way of determining which of the ions in a (alkali halide) crystal is responsible for ionic conduction is to press two crystals of MX between electrodes of the metal. (See Fig. 3.6.) If the conduction is entirely by movement of the cations then metal ions M^+ will dissolve from the positive electrode (the anode) and when they reach the negative electrode (the cathode) they will be neutralized and deposited as metal. The lattice of X^- ions remains the same. The thicknesses of the two slabs are unchanged. The cathode will grow at the expense of the anode. In the other extreme when conduction depends only on the movement of anions X^- ions will leave the region of the cathode and when they reach the anode they will react to form new layers of the salt MX. Again, the anode decreases as the cathode increases, but this time the crystal nearest the anode will increase at the expense of the crystal near the cathode. If both types of conduction are involved the result will be intermediate and the relative contributions from the different ions can be determined.

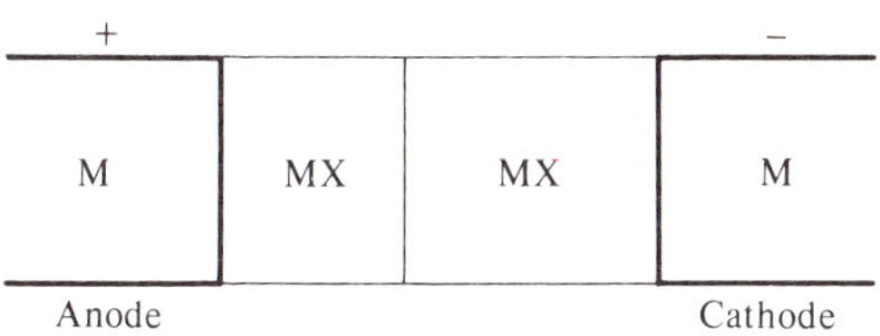

Fig. 3.6. Ionic conductivity of a salt MX.

The results of this type of experiment may also throw some light on the type of defect present. If both types of ion in the crystal MX help to conduct the current then it is unlikely that only Frenkel defects are present. Because ion size is important Frenkel defects tend to be restricted to one sublattice – usually the cation sublattice. However, conduction by only one type of ion does not mean that Schottky defects are absent: the mobility of the cations need not necessarily be the same as the mobility of the anions.

Let us suppose that we have a crystal in which, for one reason or another, we believe that Schottky defects are far more numerous than Frenkel defects and that ionic conduction is almost all due to migration of the cations. Measurement of the electrical conductivity at various temperatures should then show it to be proportional to $\exp(-U/k_BT) \times \exp(-e_S/2k_BT)$. The second exponential is simply the dependence of the number of defects on the absolute temperature. The first term is the dependence of mobility on temperature. U is known as the activation energy for ionic migration. e_S can be deduced from the experimental curves provided that U is known. The activation energy can be determined directly. One way of doing this is to deposit on one surface of a crystal a layer of the salt containing a radio-isotope. The radioactive ions will gradually diffuse through the crystal and their progress can be followed by radiation counters. Another way is to measure conductivity versus temperature again but this time to keep the number of cation vacancies artificially fixed by doping the crystal with high-charge ions (dipositive ions in the alkali halides, for example). The number of these vacancies is not dependent on temperature and if the crystal is highly doped the overall dependence of vacancy number on temperature will be unimportant. The conductivity will then be proportional to $\exp(-U/k_B T)$ and U can be found. Hence e_S can be calculated.

A more direct way of measuring e_S or e_{Fr} is by studying the way in which the specific heat increases with temperature. Energy is needed to create lattice defects and, as more lattice defects are being created per degree rise in temperature at higher temperatures than at lower ones, there will be a contribution to the specific heat which increases with temperature. Occasionally, in favourable circumstances, this contribution can be separated from other contributions to the specific heat enabling one to make an estimate of the defect formation energy.

From these, and other experiments, it is found that the alkali halides suffer mainly from Schottky defects. (e_S for sodium chloride is approximately 2 eV which means that at room temperature n_S/N is of the order of 10^{-16}.) Above about 500°C both cations and anions help to conduct electricity. Below this temperature one type of ionic conduction predominates. Sodium, potassium, and rubidium halides become cationic conductors. The caesium halides become anionic conductors. The silver halides suffer from Frenkel defects in the cationic sublattice and conduction is by diffusion of the silver ions. Compounds having the fluorite structure conduct because of Frenkel defects in the anion sublattice.

Example

1969 Paper IVA

The cohesive energy of a sodium chloride crystal per Na^+Cl^- ion pair is 6 eV. The ionization energy of Na is 5 eV and the electron affinity of Cl is 3.75 eV. Make a simple calculation to find the interionic spacing in ångström units. Indicate briefly what other factors should be taken into account in order to obtain a more accurate value of the spacing. (The electron affinity of Cl is the binding energy of the last electron in Cl^-.)

Answer

Suppose that the spacing is a_0. Then the energy needed to separate the lattice into ions (per ion pair) is

$$-\epsilon_L = \left(\frac{\alpha e^2}{4\pi\epsilon_0 a_0} - \frac{\beta}{a_0{}^n}\right)$$

where α is the Madelung constant and e is the charge on the ion. The second term inside the brackets is the repulsive term. If the cohesive energy is the energy needed to break the lattice plus the energy needed to transfer the electron from the chloride ion to the sodium ion, i.e., the energy needed to separate the lattice into atoms (per atom pair), then

$$6 = \left(\frac{\alpha e^2}{4\pi\epsilon_0 a_0} - \frac{\beta}{a_0{}^n}\right) - 1.25.$$

(N.B. If the sodium and the chlorine are well separated from one another then the lowest energy situation is when they exist as neutral atoms. Only when they are close does it become energetically feasible for electron transfer to take place.)

Now, ignoring the repulsive term,

$$\frac{\alpha e^2}{4\pi\epsilon_0 a_0} = 7.25 \text{ eV}.$$

This gives

$$\frac{a_0}{\alpha} = \frac{(1.60 \times 10^{-19})^2}{4\pi \times 8.85 \times 10^{-12}} \times \frac{1}{7.25 \times 1.60 \times 10^{-19}} \text{ metres}$$
$$\sim 2 \times 10^{-10} \text{ metres}$$

α is approximately 1.7 and therefore a_0 is of the order of 3.4 Å. The repulsive forces have been ignored. If they are to be included

$$\alpha e^2/a_0 > 7.25 \text{ eV and hence } a_0 < 3.4 \text{ Å}.$$

Questions

1) *Bristol, 1972*

Assuming that the repulsive energy between neighbouring ions in an *ionic* crystal varies as r^{-n}, where r is their separation, derive an expression for the compressibility K of the crystal. Deduce the value of n for NaCl, if

$$1/K = 2.4\ 10^{11} \text{ dyne cm}^{-2},$$

and the ionic binding energy is

$$7.6\ 10^{12}\ \text{erg mole}^{-1}.$$

(Molecular weight of NaCl: 58.5; density: 2.17 gm cm^{-2}.) Estimate the ideal tensile strength of NaCl.
[1 joule = 10^7 erg: 1 newton = 10^5 dyne: 1 a.m.u. = 1.66 10^{-27} kg.]

2) *Manchester, 1973*
Describe and discuss the nature of the forces acting on an ion in an ionic crystal such as sodium chloride.

If the repulsive part of the potential between neighbouring ions in NaCl has the form $A\ \exp(-r/\rho)$, the distance between nearest neighbours is 2.8 10^{-10} m, the Madelung constant is 1.75, and the binding energy is 763 kJ/mole, find ρ. In the light of your result, comment on any approximations made.

References

1. L. Pauling. *The Nature of the Chemical Bond*, chapter 13. Cornell UP, 1960.
2. N. N. Greenwood. *Ionic crystals, lattice defects and nonstoichiometry*. Butterworth, 1968.

4. The Free-Electron Theory of Metals

A metal is characterized by a high electrical conductivity. The electric current is carried by the 'valence' electrons which are able to move easily through the solid. How can we treat these electrons mathematically in order to explain the observed properties of metals?

In a volume of metal of one cubic centimetre there are $10^{\sim 22}$ atoms. If this metal is an alkali metal there will be one valence electron for each atom – otherwise there may be more. A strict treatment of these electrons (not counting the other electrons and the nuclei in the solid) would involve the solution of $10^{\sim 22}$ simultaneous equations. This is obviously out of the question. What we need is a simple, workable model which explains some of the physical properties. The simplest such model is the Sommerfeld model. Each valence electron is taken to be 'free'. Its movement is considered to be independent of the details of the motion of all the other electrons and the nuclei in the solid. The only effect of these other particles is to provide a potential well in which the electron moves. Sommerfeld suggested a square-well potential like that shown in Fig. 4.1. For the moment we can suppose that the walls of the potential well are infinitely high.

The allowed electron states within the metal are then solutions of the Schrödinger equation

$$\left(-\frac{\hbar^2}{2m}\nabla^2 + V\right)\psi = E\psi$$

and if we measure energy from the bottom of the well, then the states are solutions of

$$-\frac{\hbar^2}{2m}\nabla^2 \psi = E\psi$$

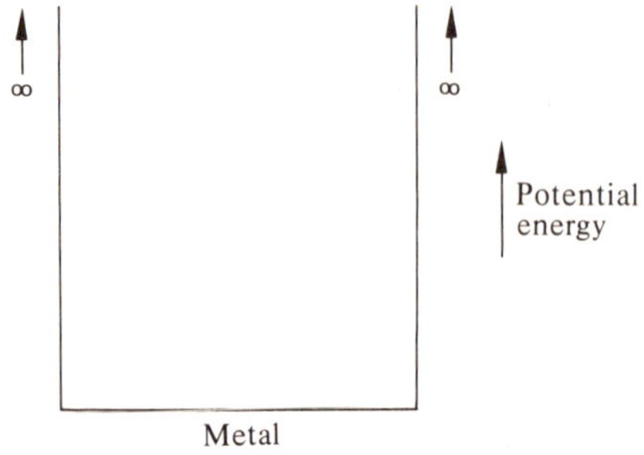

Fig. 4.1. A square-well potential for electrons in a metal.

such that $\psi = 0$ on the metal surface. The ψ's are standing waves of the form $\psi \sim \sin k_x x \cdot \sin k_y y \cdot \sin k_z z$. There is no loss of generality if we consider our piece of metal to be a cube. The number of possible states with wave vectors $k = 2\pi/\lambda$ between k and $k + dk$ can be found in exactly the same way as the number of vibrational modes of an elastic continuum was calculated on p. 46. The number of states here is

$$\frac{1}{8}\, 4\pi k^2\, dk\, \frac{V}{\pi^3} \times 2.$$

We still have the factor $^1/_8$ because only the positive octant of the sphere is considered. The last factor is now 2 corresponding to the two electron states with the same wave vector. One has 'spin up' and the other 'spin down' for the electron has a spin of ½ (angular momentum ½ħ). Now $\hbar^2 k^2/2m = E$; $\hbar k$ is the momentum of the particle. So

$$Z(E)\, dE = \frac{1}{8}\, 4\pi \left(\frac{2mE}{\hbar^2}\right)^{1/2} dE\, \frac{m}{\hbar^2}\, \frac{V}{\pi^3} \times 2$$

or

$$= CE^{1/2}\, dE \text{ where } C = \frac{V}{2\pi^2}\left(\frac{2m}{\hbar^2}\right)^{3/2}.$$

Now suppose we say that the probability of a state being occupied is given by the classical distribution function as used in chapter 2, i.e., $\exp(-E/k_B T)$. Then the average energy of an electron in the electron gas would be

$$\frac{\int_{E=0}^{\infty} E\, Z(E) \exp(-E/k_B T) dE}{\int_0^{\infty} Z(E) \exp(-E/k_B T)\, dE} = \frac{\int_0^{\infty} E^{3/2} \exp(-E/k_B T) dE}{\int_0^{\infty} E^{1/2} \exp(-E/k_B T) dE} = \frac{3}{2} k_B T$$

which is the classical expression for the energy of a monatomic gas. This would predict that the specific heat per mole of a monovalent metal should be $3Nk_B + {}^3/_2 Nk_B$, half as much again as the specific heat of an insulator. The observed electronic specific heat is actually very small and is found to depend linearly on the temperature. We must be doing something wrong. Is it that the allowed energy levels are too far apart to allow the use of a continuous function $Z(E)$? In fact this is not the problem. What is wrong is the use of $\exp(-E/k_B T)$ to express the probability of occupation of an energy level. Electrons (with a half-integral spin) obey Fermi–Dirac statistics. They are subject to the Pauli exclusion principle which forbids more than one electron to be in each of the states described above. This is a restriction which did not concern us in chapter 2. The monatomic gas may perhaps obey Fermi–Dirac statistics but at ordinary temperatures a classical distribution would not predict any appreciable probability of an energy level being more than singly occupied. Unfortunately, electrons, with their very small mass, have wider spaced energy levels and a classical distribution would contradict the Pauli principle too drastically. It is therefore necessary, as Sommerfeld realized, to replace $\exp(-E/k_B T)$ by $F(E)$ – the Fermi–Dirac distribution function – which is of the form

$$F(E) = \frac{1}{\exp\{(E - E_F)/k_B T\} + 1}$$

where E_F is known as the Fermi level. At $T = 0$ K, $F(E)$ is 1 for all values of E less than E_F and is equal to zero for all E greater than E_F. $F(E)$ is sketched in Fig. 4.2. For any temperature, $F(E)$ at $E = E_F = ½$. The definition of the Fermi level is then that it is the energy where the probability of occupation of an energy level is ½. It is easy to calculate E_{F0} – the Fermi level at absolute zero.

$$\int_0^\infty F(E)\, Z(E)\, \mathrm{d}E = NV \; (N = \text{the number of electrons per unit volume})$$

$$\int_0^{E_{F0}} 1\, CE^{1/2}\, \mathrm{d}E = C\, {}^2/_3 E_{F0}{}^{3/2} = NV$$

so

$$E_{F0} = \left(\frac{3NV}{2}\ \frac{2\pi^2}{V}\right)^{2/3} \frac{\hbar^2}{2m} = \frac{\hbar^2}{2m}\,(3\pi^2 N)^{2/3}.$$

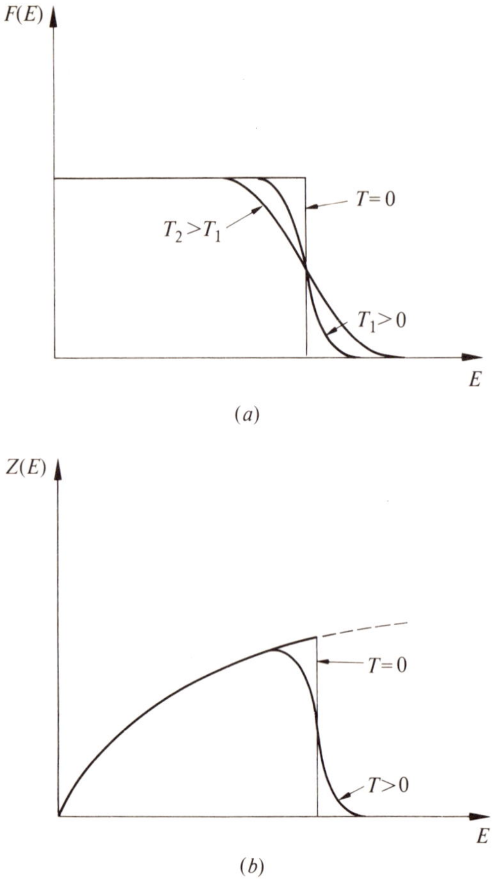

Fig. 4.2. (a) The Fermi-Dirac distribution function as a function of energy. (b) Number of available states per unit energy range as a function of energy.

Some Fermi levels (strictly at absolute zero for E_F decreases very slightly with temperature) are shown in Table 4.1. Also shown is T_F where $k_B T_F = E_F$. Only above these temperatures would the classical distribution function be useful. It can be seen that T_F is very high – in all cases far above the melting points of the metals.

TABLE 4.1

	E_{F0}/(eV) (calculated)	T_F/(K)	melting point (K)
Na	3.2	3.7×10^4	371
Cu	7.0	8.1×10^4	1356
Ag	5.5	6.4×10^4	1234

At $T = 0$ K the average energy of an electron is given by

$$\langle E_0 \rangle = \frac{1}{VN} \int_0^{E_{F0}} 1 \times C \times E^{3/2}\, dE$$

$$= \frac{1}{VN}\, C\, \frac{2}{5} E_{F0}^{5/2} = \frac{3}{5} \times \frac{2}{3}\, \frac{C E_{F0}{}^{3/2}}{VN} \times E_{F0} = \frac{3}{5} E_{F0}.$$

This is really quite a high energy – much greater than the spacing between the allowed levels in any macroscopic specimen of metal. The use of a function $Z(E)$ is valid.

The specific heat

In order to calculate the electronic specific heat we need to know $\langle E_T \rangle$ – the average energy of an electron in the metal at a temperature T. The differential of this with respect to T gives C_V.

$$NV\langle E_T \rangle = \int_0^{\infty} \frac{C E^{1/2} E}{\exp\{(E - E_F)/k_B T\} + 1}\, dE$$

and E_F is given by

$$NV = \int_0^{\infty} \frac{C E^{1/2}}{\exp\{(E - E_F)/k_B T\} + 1}\, dE\,.$$

These integrals are very difficult but the answers turn out to be

$$\langle E \rangle_T \approx \langle E_0 \rangle \left[1 + \frac{5\pi^2}{12} \left(\frac{k_B T}{E_{F0}} \right)^2 \right]$$

$$E_F \approx E_{F0} \left[1 - \frac{\pi^2}{12} \left(\frac{k_B T}{E_{F0}} \right)^2 \right].$$

The specific heat at constant volume per electron is then

$$\langle E_0 \rangle \frac{5}{6} \pi^2 \frac{k_B{}^2 T}{E_{F0}} = \frac{3}{5} E_{F0}\, \frac{5}{6}\, \frac{\pi^2 k_B{}^2 T}{E_{F0}} = \frac{\pi^2}{2}\, \frac{k_D{}^2 T}{E_{F0}}\,.$$

This can be written as

$$\frac{\pi^2}{2}\, \frac{k_B T}{T_F}$$

which, as $T \ll T_F$ at normal temperatures, is much less than the $^3/_2 k_B$ suggested by a classical free-electron model and agrees much better with experiment. The agreement is not exact because any free-electron model is an oversimplification of the real situation which is better represented by a band model – see chapter 5.

We can explain the low value of C_v as compared with a classical monatomic gas by saying that only those electrons in energy levels near to the Fermi level can be excited and contribute to the specific heat. Electrons in low energy levels cannot take small amounts of energy because all the levels immediately above theirs are already occupied. Let us say that the electrons which contribute to the specific heat are those within xk_BT of the Fermi level. This number is

$$^1/_2 (= F(E) \text{ at } E_F)\, 2xk_BT \times Z(E_F)$$

if we assume that $Z(E)$ is constant across the energy range $2xk_BT$. Let us also assume that $E_F = E_{F0}$.

$$Z(E_{F0}) = CE_{F0}{}^{1/2} = \frac{3}{2}\frac{NV}{E_{F0}}.$$

If each of these electrons contributes $3k_B/2$ to the specific heat of the electron gas then the specific heat at constant volume per electron is

$$\frac{3}{2}k_B\, xk_BT\frac{3}{2}\frac{NV}{E_{F0}}\frac{1}{NV} = \frac{9x}{4}\frac{k_B{}^2T}{E_{F0}}$$

and for this to agree with $C_v = \pi^2 k_B{}^2 T/2E_F$, x must be of the order of 2. In other words, only electrons in energy levels within about $2k_BT$ each side of the Fermi level contribute to the electronic specific heat.

Electrical and thermal conductivity

As I mentioned at the beginning of this chapter the idea of a free-electron theory of metals arose because of the high electrical conductivity of these solids. Will a free-electron model explain the observed facts? As the model stands there can be no scattering of the electrons by the ion cores, which only interact with the electrons insomuch as they contribute to the smeared-out potential, and so the conductivity would be infinite. The model must therefore be slightly modified. The energy levels are assumed to remain unchanged but the possibility of electrons being scattered by the lattice must be taken into account.

Drude produced an expression for the conductivity deduced from a classical free-electron model. He obtained

$$\sigma_C = N\frac{e^2}{m}\tau_C$$

where N is the number of electrons per unit volume, and τ_C is the relaxation time and equal to $\lambda/\bar{v}$ with λ the mean free path of electrons before collision. In his simple model Drude assumed that all the electrons had the same average velocity $\bar{v}$ before the electric field was applied. A more sophisticated treatment of the classical model by Lorentz left this result practically unchanged. (Lorentz made use of the Boltzmann transport equation – see Blakemore[1] or Dekker[2].)

Sommerfeld's calculation, using a Fermi–Dirac distribution function, gave

$$\sigma_F = N \frac{e^2}{m} \tau_F$$

but here τ_F is the relaxation time of the electrons *at the Fermi level.* All the electrons take part in the conduction process but only the relaxation time of the electrons at the Fermi level is important. Figure 4.3 may help you to understand this. The diagrams are two-dimensional representations of the allowed electron states in

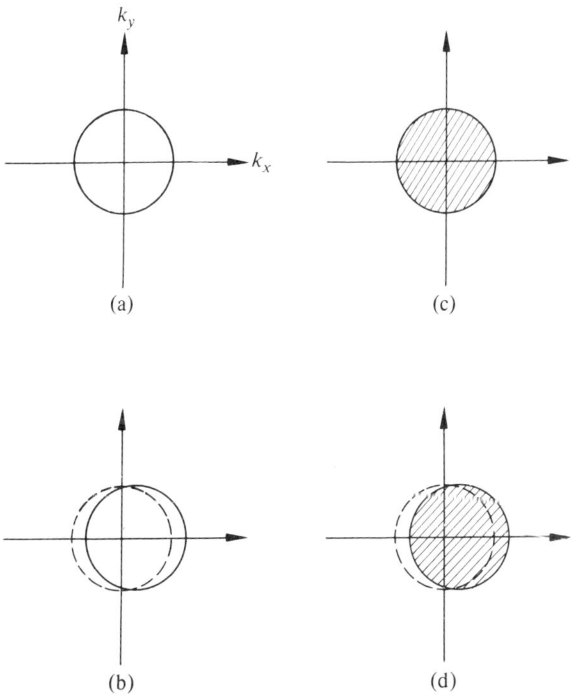

Fig. 4.3. Two-dimensional representation of Drude and Sommerfeld models of electrical conduction.

k-space. Here the allowed k's correspond to running waves rather than standing waves and so all the states with the same value of k are spread out over the whole circumference of the circle instead of just that part in the positive quadrant. (In three dimensions the allowed states with a particular k spread over the whole surface of a sphere and not just the positive octant. The total number of states is unaltered.)

Figure 4.3(a) shows the Drude distribution before a field is applied; (b) shows what happens when a field has been applied and an electric current is flowing. If the field is switched off all the electrons must be scattered before the distribution returns to (a). Part (c) shows the Fermi–Dirac distribution before a field is applied and (d) shows this distribution once a current flows. When the field is switched off only those electrons near the Fermi surface need to change their wave vectors in order that the distribution should return to that in (c).

All these calculations assume that the scattering is isotropic – that, after an electron is scattered, all directions for its new velocity are equally probable.

Experimentally, it is found that σ varies as T^{-1} at temperatures above the Debye temperature (see p. 47). We cannot tell whether σ_C or σ_F agree with experiment until we know how τ_C and τ_F depend on temperature.

A purely classical picture of electrons as hard spheres colliding with other hard spheres – the ion cores – would suggest a mean free path λ which was more or less independent of temperature. $\tau_C = \lambda/\bar{v}$ would then depend on $T^{-1/2}$ whereas $\tau_F = \lambda/v_F$ would not depend on temperature. σ_C would thus depend on $T^{-1/2}$ and σ_F would be temperature independent. *Neither* equation would agree with experiment.

There was another problem with the Sommerfeld model. When λ was calculated from experimental data it was found to be of the order of hundreds of lattice spacings. This seemed impossible until a wave-mechanical treatment of the electron scattering process showed that a perfectly periodic lattice does not scatter the electron waves. (For more about this see chapter 5.) The electrons interact only with the lattice vibrations.

An electron may interact with a phonon in such a way that the phonon is absorbed or perhaps just scattered. Electrons may also change state by emitting phonons. For any of these processes to take place the laws of conservation of energy and momentum (p. 62) must be obeyed and there must be a vacant state for the electron to be scattered into. With regard to the first of these conservation laws, it should be mentioned that the phonon energies are very small in comparison with the Fermi energy and so the electron energy changes only very slightly on scattering. The scattering of the electrons by the lattice (and hence the mean free path) is a complicated function of the temperature. At absolute zero (ignoring boundary scattering) the conductivity would be infinite if the lattice were pure.† We know that at temperatures above the Debye temperature all the vibrational modes of the metal crystal are occupied by many phonons and the energy in each mode is $k_B T$. The number of phonons in a particular mode is proportional to $k_B T$. Suppose there is an electron in a state with wave vector k which can interact with a phonon of wave vector $\boldsymbol{q}$ and go into a vacant state described by k' and that this will help to restore the distribution corresponding to zero field. The probability of the electron colliding with such a phonon is therefore proportional to the temperature. Above the Debye temperature, then, whatever the details of the scattering mechanism, the time between collisions will be inversely proportional to T.

Then $\tau \sim T^{-1}$ provided only that, at these temperatures, the conductivity is dominated by electron–phonon collisions and not by defect scattering or by processes involving phonon emission.

With $\tau \sim T^{-1}$ both equations give the correct temperature dependence. However, when Drude proposed his model he was not just using the classical distribution function instead of the Fermi–Dirac function; he was actually considering the electrons as classical particles. It would be inconsistent with such a model to use a τ which depends on a calculation making use of the quantization of vibrational energy levels and other non-classical ideas. We must have $\tau_C \sim T^{-1/2}$ to be consistent but we can have $\tau_F \sim T^{-1}$.

A Fermi–Dirac free-electron model (modified slightly to allow some interaction between the lattice and the electrons) can therefore explain the observed temperature dependence of the electrical conductivity at temperatures above the Debye temperature.

The Wiedemann–Franz law states that, above the Debye temperature, the ratio of thermal to electrical conductivity is proportional to T and that the constant of proportionality is nearly the same for all metals.

Can we explain this law? The thermal conductivity of metals is sometimes as much as a couple of orders of magnitude greater than that of insulators except at low temperatures. It is therefore a reasonable approximation to say that *all* the thermal conductivity is due to the electron gas in the metals.

†Nothing to do with superconductivity – a phenomenon which can take place in very 'dirty' lattices.

Kinetic theory gives the result that the thermal conductivity of a gas is $C_V\bar{v}l/3$ where C_V is the heat capacity per unit volume, $\bar{v}$ is the average velocity of the particles, and l is their mean free path. I shall not derive the expression but it is easy to see that it is a sensible one. As the definition of thermal conductivity is

$$\frac{\partial Q}{\partial t} = A\kappa \frac{\partial T}{\partial x}$$

with Q = heat, t = time, A = area, and T = temperature, then κ must have units of joule m^{-1} s^{-1} K^{-1}. One would expect κ to be directly proportional to C_V, $\bar{v}$ and l. Multiplying them together, all to the first power, gives the quantity $C_V\bar{v}l$ with the correct units of joule m^{-1} s^{-1} K^{-1}.

The thermal conductivity of the metal is then

$$\kappa = \frac{C_V\bar{v}l}{3}.$$

The one-third is a geometrical factor. It was shown earlier in the chapter that the average heat capacity per electron was $\pi^2 k_B{}^2T/2E_F$. E_F can be written as $\frac{1}{2}mv_F{}^2$ where v_F is the velocity of electrons at the Fermi level. It is only the electrons near the Fermi level that actually contribute to the specific heat. Only these electrons can be scattered and change their energy and so contribute to the thermal conductivity. With N electrons per unit volume

$$\kappa = \frac{1}{3} \cdot \frac{N\pi^2 k_B{}^2 T}{2(\frac{1}{2}mv_F{}^2)} \cdot v_F \cdot l$$

l is now the mean free path of electrons at the Fermi level and can be written as $v_F\tau$ where τ is the relaxation time of these electrons. The electrical conductivity

$$\sigma = N\frac{e^2}{m}\tau.$$

Therefore

$$\frac{\kappa}{\sigma} = \left(\frac{\pi^2 k_B{}^2}{3e^2}\right) T.$$

The number in brackets is known as the Lorenz number. It is independent of N and τ and the calculated value is in good agreement with experiment.

The paramagnetic susceptibility of free electrons

It is possible, on a free-electron model, to calculate the paramagnetic contribution of the conduction electrons to the magnetic susceptibility. Each electron has a spin of ½ħ units of angular momentum. Associated with the spin is a magnetic dipole moment of one Bohr magneton in the opposite direction to the spin.

When there is no magnetic field acting on a metal there are an equal number of electrons with spin dipoles 'up' and with spin dipoles 'down'. The energies of the two states are equal. But when a magnetic field B is applied, those electrons with dipoles parallel to the field have their energy changed by an amount $-\beta B$ where β is the Bohr magneton; those electrons with dipoles antiparallel to B have their energy changed by $+\beta B$. This situation is shown in Fig. 4.4(a). Such a state is obviously unstable. Some of the electrons with dipoles antiparallel will flip over and their dipoles will become parallel to the field.

The number of states of each spin at the Fermi level is

$$\tfrac{1}{2} Z(E_F)\,\mathrm{d}E = \frac{1}{2} \cdot \frac{1}{2\pi^2}\left(\frac{2m}{\hbar^2}\right)^{3/2} \cdot E_F^{1/2}\,\mathrm{d}E$$

$$= \frac{1}{2} \cdot \frac{1}{2\pi^2}\left(\frac{2m}{\hbar^2}\right)^{3/2} \cdot \left(\frac{\hbar^2}{2m}\right)^{1/2} (3\pi^2 N)^{1/3} \cdot \mathrm{d}E$$

$$= \frac{3N}{4E_F}\,\mathrm{d}E.$$

The number of electrons which change spin is $\beta B \times (3N/4E_F)$ if βB is small in comparison with E_F so that $Z(E)$ can be regarded as a constant. This gives the situation of Fig. 4.4(b). The number of parallel dipoles in excess over the antiparallel ones is just twice this, i.e., $\beta B(3N/2E_F)$. Susceptibility,

$$\chi = \frac{M}{H} = \mu_0 \frac{M}{B} \quad \text{(if } \chi \text{ is small)}$$

$$= \frac{\mu_0 \beta}{B} \cdot \beta B \cdot \frac{3N}{2E_F} = \frac{3\mu_0 \beta^2 N}{2E_F} \quad \text{(per unit volume).}$$

As the temperature increases E_F changes, but only slowly, and the paramagnetic susceptibility should be virtually independent of temperature. The theory here is moderately successful. Experimental measurements of the susceptibility include a diamagnetic contribution from the conduction electrons (for free electrons this is minus one-third of the paramagnetic contribution) and a diamagnetic contribution from the ion cores (which can be estimated from non-metallic

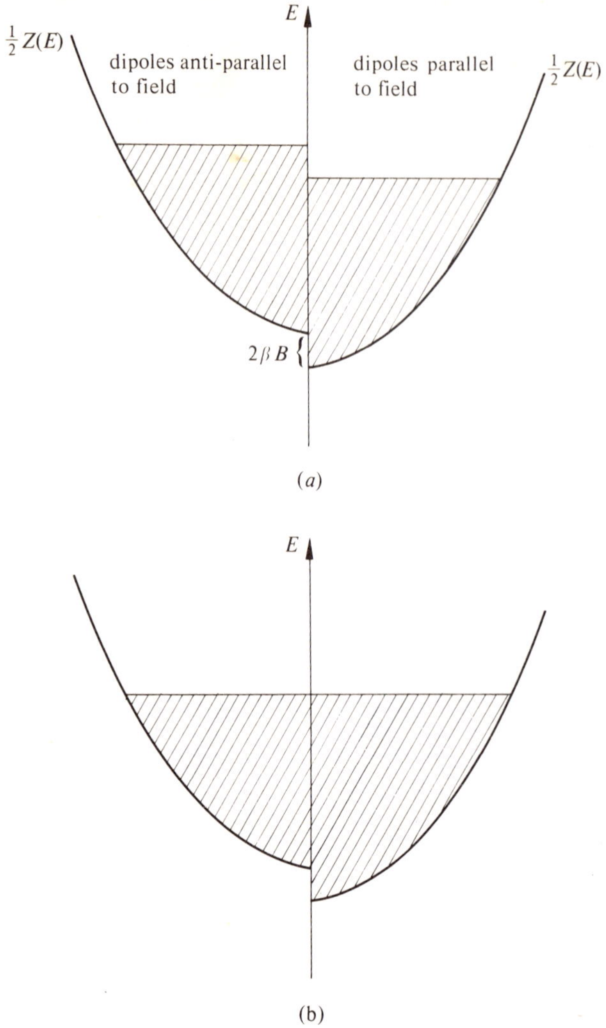

Fig. 4.4. The paramagnetism of conduction electrons.

neighbours in the periodic table). Once these other terms have been subtracted it is found that the paramagnetic contribution is indeed temperature independent and of approximately the magnitude estimated. If all the electrons were bound to their respective ions, or if they were free but obeyed classical statistics, the paramagnetic contribution would be much larger and would be temperature dependent (see chapter 6). Exact quantitative agreement between the theory and experiment is not to be expected, for our model is too simple and does not take into account the interactions of the electrons with themselves or with the ion cores.

Thermionic emission

When a piece of metal is heated, electrons are emitted from the surface. Suppose a positively charged electrode is placed near to the sample which is kept negative so that there is just enough potential difference between the sample and the electrode to draw away any emitted electrons without pulling more out of the sample. The *thermionic emission current* can then be measured as a function of temperature.

In order to attempt a calculation of this current we have to modify the potential well of Fig. 4.1. If the sides were really infinitely high then no electrons could ever get out of the metal. The potential must now be taken to be as in Fig. 4.5. (The allowed energy levels will be very slightly different for a well with finite sides but this effect can be ignored.) The Fermi level is ϕ units of energy below the vacuum level. ϕ is known as the work function. As an example of the order of ϕ, the work function of sodium is ~2.3 eV and those of copper and silver are ~4.5 eV.

Let us consider emission from a metal surface which is a plane with a normal along the x-axis. Then in order for an electron to escape from the metal, it must have

$$\frac{p_x^2}{2m} \geqslant E_F + \phi.$$

p_y and p_z can have any value. The number of electron states with $|k|$ between $|k|$ and $|k + dk| = \{2/(8\pi^3)\}\,4\pi k^2 dk$.

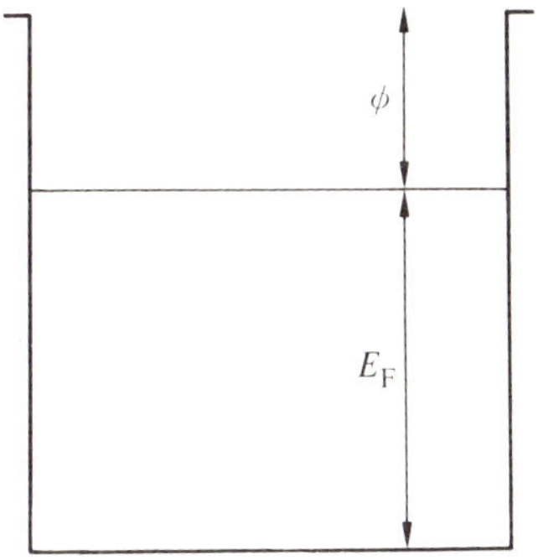

Fig. 4.5. Modified form of the potential well of Fig. 4.1. ϕ is the work function and E_F is the Fermi energy.

$$\int_{k=0}^{\infty} 4\pi k^2\,\mathrm{d}k \quad \text{can be replaced by}$$

$$\int_{k_x=-\infty}^{\infty}\int_{k_y=-\infty}^{\infty}\int_{k_z=-\infty}^{\infty} \mathrm{d}k_x\,\mathrm{d}k_y\,\mathrm{d}k_z.$$

Therefore the number of electron states per unit volume with p_x in the range $p_x \rightarrow p_x + \mathrm{d}p_x$, etc., is given by

$$\frac{2}{8\pi^3}\,\frac{\mathrm{d}p_x\,\mathrm{d}p_y\,\mathrm{d}p_z}{\hbar^3}.$$

The number of such electrons arriving at unit area of the surface per unit time is just

$$\frac{p_x}{m}\cdot\frac{2}{(2\pi\hbar)^3}\,\frac{\mathrm{d}p_x\,\mathrm{d}p_y\,\mathrm{d}p_z}{\exp[(E-E_\mathrm{F})/k_\mathrm{B}T]+1}.$$

We can write $\exp(E/k_\mathrm{B}T)$ as

$$\exp(p_x{}^2/2mk_\mathrm{B}T)\cdot\exp(p_y{}^2/2mk_\mathrm{B}T)\cdot\exp(p_z{}^2/2mk_\mathrm{B}T)$$

and if we are dealing with energies great enough for emission then $E > E_\mathrm{F}$ and we can ignore the unity in the denominator of the Fermi function. *If all* the electrons which arrive at the surface with sufficient energy are emitted,

Number of electrons emitted per unit area per second

$$= \int_{p_x=\{2m(E_\mathrm{F}+\phi)\}^{1/2}}^{\infty}\int_{p_y-\infty}^{\infty}\int_{p_z=-\infty}^{\infty}\frac{p_x}{m}\,\frac{2}{(2\pi\hbar)^3}\exp(E_\mathrm{F}/k_\mathrm{B}T)$$

$$\times\exp(-p_x{}^2/2mk_\mathrm{B}T)\exp(-p_y{}^2/2mk_\mathrm{B}T)\exp(-p_z{}^2/2mk_\mathrm{F}T)\,\mathrm{d}p_x\mathrm{d}p_y\mathrm{d}p_z$$

$$= \frac{2}{m(2\pi\hbar)^3}\exp(E_\mathrm{F}/k_\mathrm{B}T)\int_{p_y=-\infty}^{\infty}\exp(-p_y{}^2)/2mk_\mathrm{B}T)\,\mathrm{d}p_y$$

$$\int_{p_z=-\infty}^{\infty}\exp(-p_z{}^2/2mk_\mathrm{B}T)\,\mathrm{d}p_z$$

$$\times\int_{\{p_x=2m(E_\mathrm{F}+\phi)\}^{1/2}}^{\infty} p_x\exp(-p_x{}^2/2mk_\mathrm{B}T)\,\mathrm{d}p_x.$$

The first two integrals are gamma functions (see footnote on p. 124) and are each equal to $(2\pi mk_\mathrm{B}T)^{1/2}$. The third integral is equal to

$mk_BT \exp\{(-E_F + \phi)/k_BT\}$. The thermionic emission current is given by

$$e \cdot \frac{2}{m(2\pi\hbar)^3} \exp(E_F/k_BT) \cdot 2\pi mk_BT \cdot mk_BT \exp\{-(E_F + \phi)/k_BT\}$$

$$= \frac{4\pi mek_B{}^2}{(2\pi\hbar)^3} T^2 \exp(-\phi/k_BT)$$

or

$$AT^2 \exp(-\phi/k_BT)$$

where A has the value 1.2×10^6 A m^{-2} K^{-2}.

This is known as Richardson's equation. It is in fairly good agreement with experiment provided one remembers that

1. some electrons may be reflected back at the metal surface even if their energy is greater than $E_F + \phi$,
2. ϕ is not a constant: it varies with temperature, it may vary with crystal direction, and it is very sensitive to impurities on the metal surface.

Contact potentials

It is known that there is a potential difference between two dissimilar metals in contact. With the help of our modified potential we can understand what is going on.

Two metals A and B with work functions ϕ_A and ϕ_B are brought into contact (Fig. 4.6(a)). ϕ_A is less than ϕ_B. This means that some electrons in metal A have higher energies than empty states in metal B. The electrons will begin to flow across. The result of this is that metal A gains a positive charge and metal B a negative charge. The vacuum levels of the two metals are no longer equal for there is a potential gradient between them as shown in (b). Electron flow will stop when the Fermi levels of the two metals are equal – there will then be no more tendency for electrons to cross from A to B than for them to cross in the opposite direction. Very few electrons are transferred and so the potential difference at equilibrium is practically $\phi_B - \phi_A$.

Work functions, and hence contact potentials, are very sensitive to the state of the surface. Very small amounts of chemical impurities absorbed into the surface can significantly change an experimental result. Experiments must therefore be done with chemically clean surfaces in a vacuum.

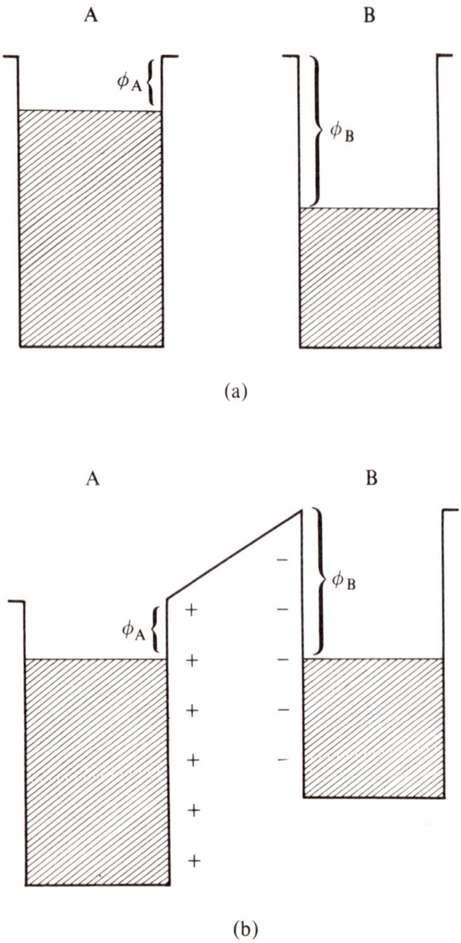

(b)

Fig. 4.6. The contact potential.

Summary
The original free-electron model had some success in explaining the order of magnitude of the electronic specific heat. Detailed comparison with experiment indicates that the theory is not quite right if E_F is calculated using the free-electron mass m. Better agreement between theory and experiment requires the use of an 'effective mass' greater than m. The theory of paramagnetic susceptibility also improves with the use of an 'effective mass.'

But these are fairly minor adjustments. The real problem is that there are some things that a free-electron model cannot even begin to explain. For example, it gives us no idea of the cohesive energy of a metal or of the elastic constants. We can show that E_F increases as the metal is compressed but this is meaningless unless we know how the bottom of the well, our zero of energy, changes in comparison with the vacuum level. The model tells us nothing about this.

There are many other problems with a free-electron model, some of which will be explained in the next chapter. Also in chapter 5, you will find how the model can be improved and the significance of 'effective mass'.

Examples

1) 1969 Option 3

A metal contains 5×10^{22} free electrons per cm^3. Estimate the contact potential which would exist between two samples of this metal if one were subjected to a pressure of 30 atmospheres and the other maintained at atmospheric pressure, assuming that the bulk modulus is 10^{12} dyne $cm^{-2} = 10^{11}$ newton m^{-2}. (1 atmos. = 10^6 dyne $cm^{-2} = 10^5$ newton m^{-2}.)

Answer

Bulk modulus

$$B = -V\frac{\partial P}{\partial V} .$$

If $B = 10^{11}$ and $\delta P = 29 \times 10^5$, then

$$-\frac{\delta V}{V} = \frac{29 \times 10^5}{10^{11}} = 2.9 \times 10^{-5} .$$

The number of electrons per unit volume will also change.

$$nV = (n + \delta n)(V + \delta V)$$

so

$$n + \delta n = \frac{nV}{V + \delta V} = \frac{n}{1 - 2.9 \times 10^{-5}}$$

and

$$\delta n = (5 \times 10^{28})(2.9 \times 10^{-5}). \quad (n = 5 \times 10^{28} \text{ per } metre^3.)$$

The expression for the Fermi energy is

$$E_F = \frac{\hbar^2}{2m}(3\pi^2 n)^{2/3}$$

and so the change in the Fermi energy can be written as

$$\frac{\delta E_F}{\delta n} = \frac{\hbar^2}{2m}(3\pi^2)^{2/3}\frac{2}{3}n^{-1/3} \quad \left(\text{for } \frac{\delta n}{n} \text{ small}\right).$$

$$\delta E_F = \frac{(1.05 \times 10^{-34})^2}{2 \times 9.11 \times 10^{-31}}(3\pi^2)^{2/3}\frac{2}{3}(5 \times 10^{28})^{2/3}(2.9 \times 10^{-5}) \text{ joules}$$

$$\sim 1.5 \times 10^{-23} \text{ joules} = \sim 1 \times 10^{-4} \text{ eV}.$$

With a free-electron model we have to equate the change of Fermi energy with the contact potential. We have no means of calculating any change with pressure of the depth of the potential well.

2) 1970 Paper IVA
Estimate, using the free-electron theory, the paramagnetic contribution of the conduction electrons to the susceptibility per unit volume of potassium, for which the Fermi energy is 2.1 eV and the density of conduction electrons is 1.3×10^{22} cm^{-3}.

Answer

$$\chi = \frac{3\mu_0\beta^2 N}{2E_F}$$

$$= \frac{3}{2}\,\frac{(4\pi \times 10^{-7}) \times (9.27 \times 10^{-24})^2 \times (1.3 \times 10^{28})}{2.1 \times 1.60 \times 10^{-19}}$$

$$\sim 6 \times 10^{-6} \text{ SI units.}$$

Questions

1) *Manchester, 1974*
Discuss the application of quantum mechanics to the free-electron theory of metals mentioning in particular the eigenstates of a three-dimensional box and the occupation of those states.

A simple metal has a free electron density of 0.1 electrons per cubic ångström. What is the energy in electronvolts of an electron on the Fermi surface?

2) *Manchester, 1972*
Describe and contrast the mechanism of electrical and thermal conduction in a metal using a free-electron model.

The free-electron concentration in gold is 5.90 10^{22} electrons cm^{-3}. At 300 K the conductivity of pure gold is 5.0 10^5 Ω^{-1} cm^{-1}. Calculate the electron mean free path for phonon scattering.

3) *Manchester, 1971*
Aluminium is trivalent and crystallizes into a fcc metal. The side of the conventional fcc unit cell is 4 Å.
(a) Give a possible set of primitive lattice vectors and calculate the nearest-neighbour separation.
(b) Calculate the width of the energy region occupied by conduction electrons on the basis of the free-electron model. You may assume that the density of allowed k-values, in a crystal of volume V, is $V/(2\pi)^3$ per unit volume of k-space.

References

1. J. S. Blakemore. *Solid State Physics*. Saunders, 1970.
2. A. J. Dekker. *Solid State Physics*. Macmillan, London, 1958.

5. Band Theory and Semiconductors

The need for band theory

There are quite a number of the properties of solids which cannot possibly be explained on the basis of a free-electron model. In the last chapter you saw that some of the properties of *metals* can be explained on such a model with at least a reasonable degree of success. The conductivity of metals is one of these: but even here we have some problems – how to explain the interaction of the electron waves with the lattice.

Another problem, not mentioned in chapter 4, is the Hall coefficient of the alkaline earth metals. Details of the Hall effect can be found in, for example, Bleaney and Bleaney.[1] Basically what happens is that when a magnetic field is applied to a conductor, at right angles to the current direction, then a potential is built up across the conductor as shown in Fig. 5.1. If the current is carried by negative charges then the Hall coefficient (defined as minus the ratio of $\boldsymbol{E}$ to $\boldsymbol{I} \times \boldsymbol{B}$) will be negative as illustrated in (a). If the current is carried by positive charges, $\boldsymbol{E}$ will be as shown in (b) and the sign of the Hall coefficient will be positive. A free-electron model predicts behaviour as in (a) but experiments show that the Hall coefficient for the alkaline earth metals is positive. The current appears to be carried by positive charges!

A free-electron model will not suffice to explain all the properties of metallic solids and, as one might expect, it is even less useful for predicting the properties of semiconductors or of insulators. Semiconductors, for instance, not only frequently have Hall coefficients which indicate the movement of positive charges, but the conductivity of these materials increases exponentially with temperature. This behaviour is very unlike that of metals, which show a decreasing conductivity with increasing temperature.

Nor can the free-electron model explain why some insulators can

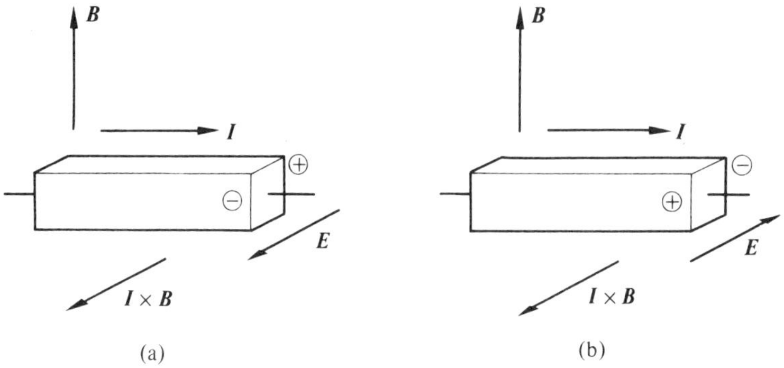

Fig. 5.1. The Hall effect (a) for negative carriers, (b) for positive carriers. *I* is the current density and is equal to the number of charge carriers per unit volume × charge × velocity.

become good conductors in the presence of electromagnetic radiation – the photoconductive effect. The only way that a free-electron model can explain poor conductivity is by poor electron mobility ($\mu = |e/m|\tau$), although even the concept of a non-infinite mobility is somewhat inconsistent with the model. But why should radiation increase mobility?

Also, as explained at the end of the last chapter, the free-electron model is unable to account in any way for the cohesive energy or elastic constants of a solid.

The Bloch functions

Hopefully we might be able to improve on the simple model of chapter 4. To include the electron correlation is still out of the question but we can at least try to improve on the square-well potential and substitute instead something which looks more like an array of atoms. The potential (one dimensional) in Fig. 5.2(b) is a bit more realistic than its counterpart in Fig. 5.2(a). The solution for these, more realistic, potentials is a formidable task. (Remember that, although we may study one-dimensional potentials for simplicity, theoreticians are solving three-dimensional problems.) There are many ways of setting about the actual solutions but there are restrictions upon the possible solutions of the Schrödinger equation which must always apply.

The potential energy V is now no longer a constant but is a function $V(x)$ which has the periodicity of the lattice. If the lattice spacing is a, then

$$V(x + a) = V(x). \tag{5.1}$$

(a)

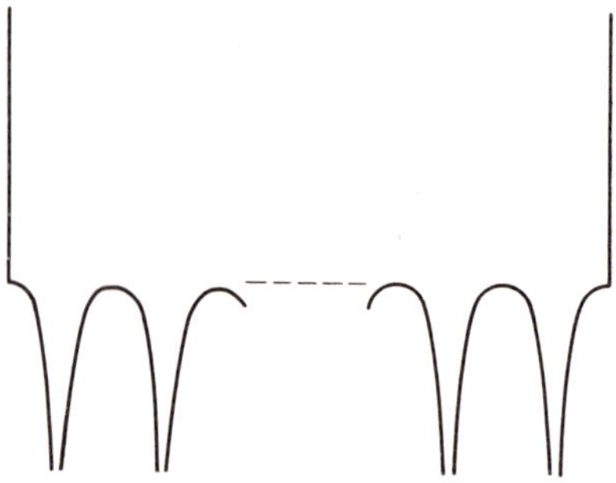

(b)

Fig. 5.2. (a) Potential used for the free-electron model. (b) A potential that could be used for a band model.

The technique of Fourier analysis is used to write this function in the form

$$V(x) = \sum_G V_G \cdot \exp(\mathrm{i}Gx). \tag{5.2}$$

If eq. (5.2) is to obey the condition of (5.1) then

$$V(x+a) = \sum_G V_G \exp\{\mathrm{i}G(x+a)\} = \sum_G V_G \exp(\mathrm{i}Gx)$$

so $\exp(\mathrm{i}Ga)$ must equal unity. This will be true if

$$Ga = 2\pi \times \text{an integer},\ G = \frac{2\pi}{a}n \quad (n = 0, \pm 1, \pm 2, \text{etc.}).$$

The G's are the reciprocal lattice vectors of a one-dimensional lattice. (In three dimensions $V(r)$ can be written as

$$\sum_G V_G \exp(\mathrm{i}G \cdot r).)$$

So we have

$$V(x) = \sum_{G=\frac{2\pi n}{a}} V_G \exp(\mathrm{i}Gx) \quad (n = 0, \pm 1, \pm 2, \text{etc.})$$

which is an infinite polynomial. If $V_G = +V_{-G}$ then the two terms add together to give $2V_G \cos Gx$ because $\exp(\mathrm{i}Gx) = \cos Gx + \mathrm{i} \sin Gx$. If all $V_G = +V_{-G}$ then $V(x)$ is not only periodic but is symmetrical about $x = 0$. If the electron wave functions which are solutions of the Schrödinger equation are called ψ, then

$$\left\{-\frac{\hbar^2}{2m}\frac{\partial^2}{\partial x^2} + V(x)\right\}\psi(x) = E\psi(x).$$

This can be written

$$\left\{-\frac{\hbar^2}{2m}\frac{\partial^2}{\partial x^2} + \sum_G V_G \exp(\mathrm{i}Gx)\right\}\psi(x) = E\psi(x). \tag{5.3}$$

The solutions of such an equation are of a special form. The *Bloch theorem* states that the eigenfunctions of such a wave equation are of the form

$$\psi_k(x) = \exp(\mathrm{i}kx)u_k(x)$$

(in three dimensions $\psi_k(r) = \exp(\mathrm{i}k \cdot r)u_k(r)$)

where $u_k(x)$ is a function with the periodicity of the potential $V(x)$ and $\psi_k(x)$ is known as a Bloch function.

Suppose that we consider a linear lattice of N 'atoms' (N even). Then, using cyclic boundary conditions (see p. 54) we have

$$\psi_k(x) = \exp(\mathrm{i}kx)\,u_k(x) = \psi_k(x + Na) = \exp\{\mathrm{i}k(x + Na)\}u_k(x + Na).$$

Now $u_k(x + Na)$ is, by definition, equal to $u_k(x)$ because we have said that u_k has the periodicity of the lattice. We must have

$$\exp(\mathrm{i}kNa) = \exp(\mathrm{i}2\pi n)$$

so

$$kNa = 2\pi n,\ k = \frac{2\pi n}{Na} \quad (n = 0, \pm 1, \pm 2, \text{etc.}).$$

When $n = N/2$ we have got to $k = \pi/a$ which is the edge of the first Brillouin zone. If we say that

$$k = \frac{2\pi n}{Na} \quad (n = 0, \pm 1, \pm 2, \ldots +N/2)$$

not counting $-N/2$ as distinct from $N/2$, then in the first Brillouin zone we have as many electron states (excluding spin) as there are atoms in the solid. If we associate two spin states with each k then the number of distinct states, corresponding to particular u_k, in the first zone is $2N$.

The Bloch functions are eigenfunctions of a perfect periodic potential. An electron in a Bloch state cannot be scattered into another Bloch state by this potential. Electrons will not be scattered by a *perfect*, infinite, periodic lattice.† You may think of it like this. If an electron (in a Bloch state) has managed to get to the lattice position under consideration, having come all the way from infinity, without being scattered, then there is no reason why it should be scattered now. Its wavefunction will be adjusted to the periodic potential. However, if the periodicity is broken – if the lattice point moves (thermal motion) or if the potential around it is different than it was in the rest of the crystal (impurity) – then the electron may be scattered provided that the laws of conservation of energy and 'momentum' can be satisfied.

How do we set about solving eq. (5.3)? The potential energy $V(x)$ of one electron depends not only on the nuclei in the lattice but also upon the distribution of all the other electrons. Assuming that the nuclei are fixed then the way to get self-consistent solutions of this equation would be as follows. First we would have to guess a function for V, say V_1, and then solve the Schrödinger equation of (5.3) to find the eigenfunctions ψ_1. However, we would then probably find that our original guess for V was inconsistent with our calculated ψ_1. This would mean using ψ_1 to give a new expression V_2 and recalculating the ψ's to find ψ_2, etc., until self-consistency was obtained. Such a task would be formidable if not impossible. However, there are situations where we have a very good idea of the form of ψ and although our guess will not be an exact eigenfunction it will be good enough to give us an idea of the allowed electron energy levels.

The nearly-free-electron approximation

We know that the behaviour of the conduction electrons in metals is reasonably well explained by a free-electron model. Such a model assumes that the electron waves are of the form

$$\psi_k = \exp(ikx) \quad \text{(times a spin part which has two values),}$$

$$k = \frac{2\pi n}{Na} \quad (n = 0, \pm 1, \ldots N/2).$$

†In the absence of an applied field. An electric field may cause back-scattering of electrons in states near the zone boundary – see later.

We would expect this to be a reasonable approximation to the true wavefunction. It is assumed, therefore, that the potential field, whatever it may happen to be, can be treated as a perturbation upon the free-electron states, i.e., plane waves. For small values of k the electron energy is to all intents the same as that of the free-electron model. But simple perturbation theory breaks down when k lies near a zone boundary. Then one has to modify the plane waves and write

$$\psi_k = \sum_G \alpha_{k-G} \exp\{\mathrm{i}(k - G)x\}. \tag{5.4}$$

If $\alpha_{k-G} = 0$ unless $G = 0$ we have a plane wave.

Is the ψ_k of eq. (5.4) a Bloch function? We can write

$$\psi_k(x) = \exp(\mathrm{i}kx) \sum_G \alpha_{k-G} \exp(-\mathrm{i}Gx)$$

and so

$$\sum_G \alpha_{k-G} \exp(-\mathrm{i}Gx)$$

must be a function with the periodicity of the lattice for $\psi_k(x)$ to be a Bloch function.

$$u_k(x + a) = \sum_G \alpha_{k-G} \exp(-\mathrm{i}Ga) \exp(-\mathrm{i}Gx).$$

But G is a reciprocal lattice vector and a is a lattice vector: $Ga = 2\pi n$ and $u_k(x + a) = u_k(x)$. Equation (5.4) obeys the Bloch condition.

When $|k| = |k - G|$ the coefficient of the wave $\exp\{\mathrm{i}(k - G)x\}$ is as important as the coefficient of $\exp(\mathrm{i}kx)$. In effect there has been Bragg refraction of the electron wave by the lattice just as if it were an external electron beam. If we write

$$\psi_k = \exp(\mathrm{i}kx) \pm \exp\{\mathrm{i}(k - G)x\}$$

then

$$E\pm = \frac{\int[\exp(-\mathrm{i}kx) \pm \exp\{-\mathrm{i}(k - G_1)x\}]\left[\dfrac{-\hbar^2}{2m}\dfrac{\partial^2}{\partial x^2} + \sum_G V_G \exp(\mathrm{i}Gx)\right] \times [\exp(\mathrm{i}kx) \pm \exp\{\mathrm{i}(k - G_1)x\}]\,\mathrm{d}x}{\int[\exp(-\mathrm{i}kx) \pm \exp\{-\mathrm{i}(k - G_1)x\}][\exp(\mathrm{i}kx) \pm \exp\{\mathrm{i}(k - G_1)x\}]\,\mathrm{d}x}$$

where the integration must be taken over a large range of values of x. Because

$$\frac{\int\exp(-\mathrm{i}Bx)\exp(\mathrm{i}Ax)\,\mathrm{d}x}{\int\exp(-\mathrm{i}Ax)\exp(\mathrm{i}Ax)\,\mathrm{d}x} \quad (A \neq B) \to 0 \text{ as the range of integration} \to \infty,$$

then a term like

$$\frac{\int \exp(-ikx)\left\{\frac{-\hbar^2}{2m}\frac{\partial^2}{\partial x^2}+\sum_G V_G \exp(iGx)\right\}\exp\{i(k-G_1)x\}\,dx}{\int[\exp(-ikx)+\exp\{-i(k-G_1)x\}][\exp(ikx)+\exp\{i(k-G_1)x\}]\,dx}$$

is equal to $V_{G_1}/2$ because all the other Fourier coefficients are multiplied by terms which go to zero on integration. It turns out that

$$E_+ - E_- = V_{G_1} + V_{G_1} = 2V_G$$

with $V(x)$ symmetric about $x = 0$. If $G_1 = 2\pi/a$, $\psi_+ = 2\cos(\pi x/a)$ and this corresponds to a build-up of charge on the lattice points. $\psi_- = 2i\sin(\pi x/a)$ and this corresponds to a build-up of charge between the lattice points. One expects E_+ to be lower than E_- because of the attractive charge at the atoms. V_{G_1} is in fact negative. This can be found from the Fourier analysis but is probably fairly easy to accept. If we try to fit a curve of the form $2V_{G_1}(\cos 2\pi x/a)$ to a potential similar to that sketched in Fig. 5.2(b) with $x = 0$ at a lattice site, then we need a negative value of V_{G_1} to reproduce the dip in the potential at $x = 0$, $x = a$, etc. There are now two allowed energies corresponding to k on the zone boundary. They are $E_0 \pm 2\,|V_{G_1}|$ where E_0 is the free electron energy, i.e., $\hbar^2 k^2/2m$. See Fig. 5.3. In one dimension the bands never overlap.

Each band corresponds to values of $k = 2\pi n/Na$ ($n = 0, 1, 2, \ldots N/2$) and so if we have N atoms in the chain then there are N states of each spin in each energy band. The alkaline earth metals – beryllium, magnesium, calcium, etc. – have two s-electrons outside a closed shell. We should then expect (by analogy with our one-dimensional model) that in a crystal of N atoms, the $2N$ electrons would just fill the lowest energy band. As will be explained later, a solid with its energy bands either full or empty rather than partially filled would be expected to be an insulator, or perhaps a semiconductor, but certainly not to behave like a metal. The problem here is that there is not a complete analogy between the one- and three-dimensional situations. In three dimensions bands can overlap so that there will be states available into which the s-electrons can be excited without having to cross an energy gap.

The tightly-bound approximation

When electrons are in states deep within the atom cores we expect the wavefunctions near the nuclei to be very similar to atomic wavefunctions. We can write the ψs

$$\psi_k(r) = \sum_l \exp(ik\cdot l)\,\phi_a(r-l) \qquad (5.5)$$

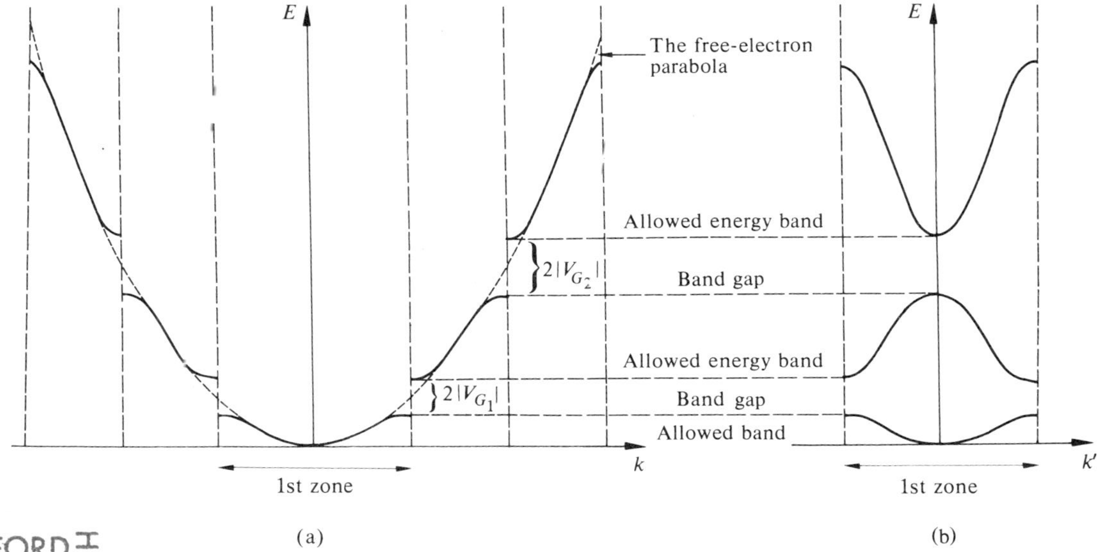

Fig. 5.3. Energy bands on the n.f.e. approximation: (a) the extended-zone scheme, (b) the reduced-zone scheme.

where $\phi_a(r - l)$ is an atomic wavefunction centred on the lattice side at l. This can be written as

$$\psi_k(r) = \exp(ik \cdot r) \sum_l \exp(\{ik \cdot (l - r)\} \phi_a(r - l).$$

When $r \to r + L$ where L is any lattice vector

$$\sum_l \exp\{ik \cdot (l - r)\} \phi_a(r - l)$$
$$\to \sum_l \exp[ik \cdot \{(l - L) - r\}] \phi_a\{r - (l - L)\}$$

which is unchanged because the sum is over all lattice vectors. Equation (5.5) satisfies the Bloch condition. Note that the Bloch functions spread throughout the crystal. An electron in a Bloch function has an equal chance of being in orbit about any of the nuclei.

For the actual calculation of the energies, V is needed. There will be contributions to the potential energy from the nuclei, from all 'tightly-bound' electrons, and also from other electrons in the solid. They may perhaps be conduction electrons or electrons in states which cannot be treated this way. The calculations are more complicated than those for the n.f.e. approximation and the details of the band shape depend strongly on the way the ϕs behave outside the core. This is not just because of the contribution that these parts of ϕ make to V but because if there were no overlap of the ϕs at all (if wherever $\phi(r - l_1)$ was non-zero, $\phi(r - l_2)$ were zero for all $l_2 \neq l_1$) then the energy *bands* would not exist. We would have the narrow energy lines of electrons in atoms, for all states would have the same energy. Superficially the bands are similar to those of the n.f.e. approximation. Here k can only vary between $-\pi/a$ and $+\pi/a$ because increasing k by a reciprocal lattice leaves ψ unchanged. Other bands can be obtained by changing the atomic orbital ϕ_a.

Both these approximations *are* only approximations. The tight-binding approximation is only really valid for narrow low-energy bands while the n.f.e. approximation can only be applied with any real hope of success to the alkali metals.

There are many variations and improvements on the above methods. When the overlap of atomic orbitals is appreciable, ϕ_a can be replaced by a linear combination of atomic orbitals, or by a wavefunction calculated using one of the cellular methods, in an attempt to produce a wavefunction which is slightly nearer to the actual solution of the Schrödinger equation. All the calculations are different and on the whole they do not produce terribly good quantitative agreement with experiment. The main point is that all the models do show that energy bands exist in solids. A great deal can be understood once this fact is appreciated even though the exact shape of the bands in k-space may be unknown.

Density of states, and constant-energy surfaces

On a free-electron model the constant-energy surfaces in three-dimensional k-space are simply spheres and the density of states $Z(E)$ is of the form $CE^{1/2}$. On a band model, the constant-energy surfaces are no longer spheres – or circles on a two-dimensional representation – but are distorted at the zone boundary. Figures 5.4(a) and (b) show

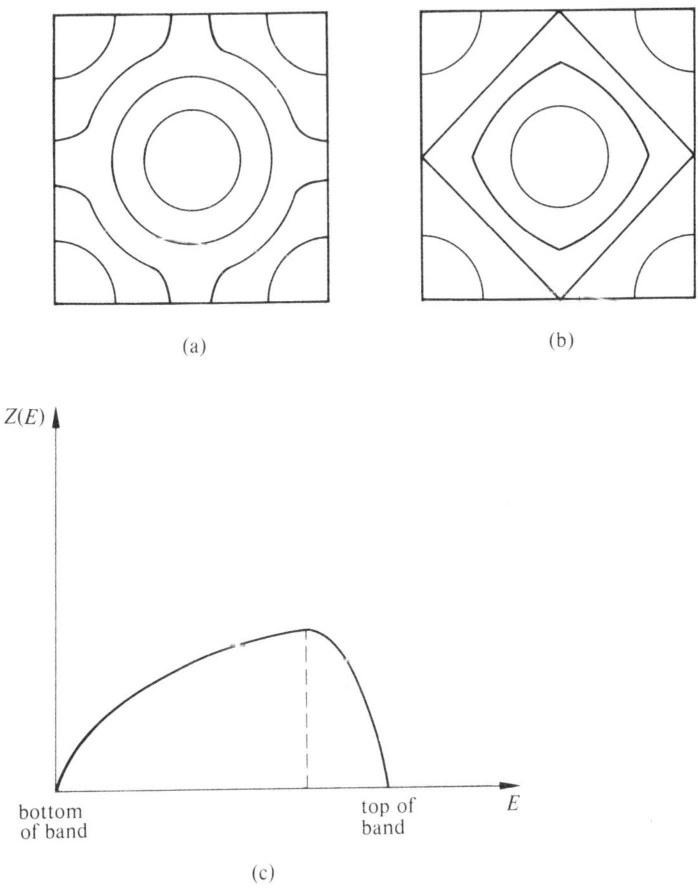

Fig. 5.4. (a) Constant-energy surfaces on a n.f.e. approximation (simple cubic lattice). (b) Constant-energy surfaces on tight-binding approximation. (c) Density of states in a band.

these contours for a n.f.e. approximation and for a tight-binding approximation. The free-electron circles continue from the origin nearly to the zone boundary in the first case but in the second case the distortion is more noticeable at smaller $|\boldsymbol{k}|$ values.

When we have bands of energy then the density of states must be zero at the top of the band as well as at the bottom and a *very simplified* $Z(E)$ is sketched in Fig. 5.4(c). The $Z(E)$ versus E curves can be obtained experimentally. X-rays are used to knock electrons out of very narrow, low-lying bands. The spectrum of soft X-rays, emitted as the electrons in higher bands make transitions to the vacant levels, is analysed to give information about the energy distribution of the higher bands.

Effective mass and the positive hole

The curve of energy against wave vector for a free-electron model is shown as a dotted line in Fig. 5.3. The equation of the curve is

$$E = \frac{\hbar^2 k^2}{2m} \tag{5.6}$$

where m is the mass of the electron.

$$\frac{\partial E}{\partial k} = \frac{\hbar^2 k}{m}, \quad \frac{\partial^2 E}{\partial k^2} = \frac{\hbar^2}{m}, \quad \text{or} \quad m = \frac{\hbar^2}{\partial^2 E/\partial k^2}$$

Now look at the solid curve of Fig. 5.3. which represents a plot of energy against wave vector for a n.f.e. model. $\partial^2 E/\partial k^2$ is not a constant as it was for the dotted parabola. As k approaches the zone boundary $\partial^2 E/\partial k^2$ first reduces to zero and then becomes negative. The effective mass m^* is now a function of k,

$$m^* = \frac{\hbar^2}{\partial^2 E/\partial k^2}$$

it can be infinite and it can be negative (see Fig. 5.5).

This should not be too worrying. It is not that the mass of the electrons in the solid changes in some mysterious manner. It is just that, *if* we insist on trying to describe as free, electrons which are in fact interacting strongly with the lattice, then in order to explain their properties using an equation like eq. (5.6) we have to adjust the parameter m and call it an effective mass.

A large or even infinite effective mass is not too difficult to accept. If there is a strong binding force between the electron and the lattice the electron will obviously be difficult to move. But why should the effective mass be negative near the zone boundary? The following

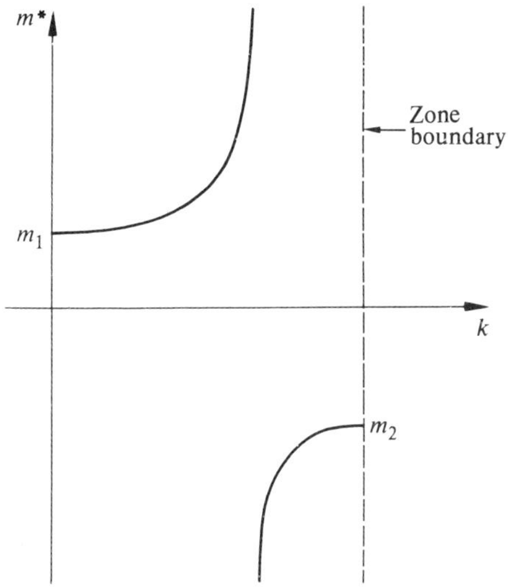

Fig. 5.5. The effective mass of an electron in an energy band. In a general case $m_1 \neq m_2 \neq m$.

simplified picture may help you to understand what is going on. Imagine an electron with a k just less than the value at the boundary. It will manage to move through the crystal. But then suppose that a field is applied which should accelerate it and increase k. As the electron responds to the field it will meet the condition for Bragg refraction and will be scattered back in the opposite direction. It will behave as would a particle with a negative charge and a negative mass. (Strictly speaking the situation in a real crystal is more complicated than this. The bands are shapes in k-space and m is a 3 by 3 tensor. A force applied in one direction may cause acceleration in another – not necessarily just the opposite – direction.)

When a band of energy levels is full apart from a few states near the top of the band, then we can say that there is a full band plus an absence of particles with negative charge and negative effective mass. An absence of particles with a negative charge and negative mass is the same as the presence of particles with positive charge and positive mass. These 'positive holes' can be excited to different energy levels and different k-values by rearrangement of the occupation of the electron states in the nearly filled band. A hole near the top of the band represents a situation of lower energy than does a hole in the electron states lower in the band. The hole energies must be measured downwards from the top of the band. A current carried by a nearly filled energy band will appear to be carried by positively charged holes moving in the current direction rather than by negatively charged electrons moving in the opposite direction. The Hall effect will not be the same as it is for the alkali metals where the current is carried by electrons in the bottom half of an energy band. Such electrons have a positive effective mass of the same order as m.

Cohesive energy and the elastic constants

Metals are bound together because the energy of the conduction electrons in the metal is lower than the orbital energy of these electrons in the free atom. In order to find out something about the cohesive energy we need to know how the energy of the band at $k = 0$, the potential energy, depends on the atomic separation, a. The n.f.e. approximation is no good to us here. It tells us nothing about the actual energy at $k = 0$ for the lattice is assumed to have no effect until $k = 2\pi/a$. It is necessary to use a Wigner–Seitz cellular method. The crystal is considered divided into small polyhedra, each centred on an atomic position. The potential inside each cell is assumed to be just that of a positive ion (with a single charge for alkali metals) at the centre of the cell.

The wave equation inside the cell is solved with the Wigner–Seitz boundary condition that $\partial\psi/\partial r$ must be zero at the boundary of the cell for the $k = 0$ state. Unlike the n.f.e. approximation where $\psi(k = 0)$ is a constant everywhere, ψ is not constant over all the cell volume. The energy $E_{k=0}$ with relation to the vacuum level can be calculated. Also the variation of $E_{k=0}$ with the size of the cell, i.e., with the interatomic spacing a, can be worked out.

The average total energy of the electrons within the solid can be written as

$$E = E_{k=0} + \langle E_{\text{kin}} \rangle.$$

If we confine ourselves to the alkali metals, we can use free-electron theory to give us a reasonably accurate expression for $\langle E_{\text{kin}} \rangle$, the average kinetic energy.

$$\langle E_{\text{kin}} \rangle = \frac{3}{5} E_F = \frac{3}{5} \frac{\hbar^2}{2m} (3\pi^2 N)^{2/3}$$

i.e., it depends on the number of electrons per unit volume. As a decreases, E_F increases.

$$E = E_{k=0} + \frac{3}{5} E_F.$$

When E is plotted against a we get a minimum at a value of a equal to a_0. a_0 is the equilibrium lattice spacing. The energy needed to change a by a small amount from a_0 indicates the elastic constants. The cohesive energy is given by

$$E_{\text{coh}} = -\left(E_{k=0} + \frac{3}{5} E_{\text{F}} + E_i\right)$$

where E_{i} is the energy needed to ionize a free atom, i.e., minus the energy of the free-atom orbital. E_{i} and E_{F} are positive; $E_{k=0}$ is negative; E_{coh} is defined as positive when the solid state is of lower energy than the atomic state.

Metals, insulators, and semiconductors

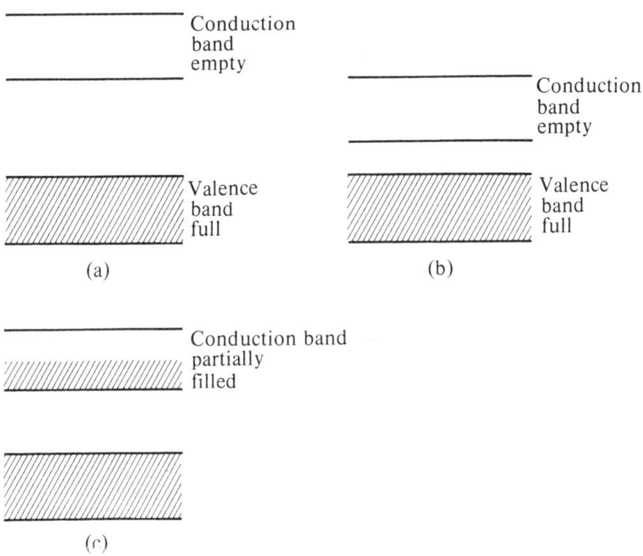

Fig. 5.6. The energy bands of (a) an insulator, (b) a semiconductor, (c) a metal, at 0 K.

Figure 5.6(a) shows the energy bands of an insulator at absolute zero. One or more bands, the valence bands, are full and the conduction band is empty. When a *small* field is applied to the solid there is no flow of current. This is *not* because the electrons in the valence band are each tied to an individual atom and cannot move, but because the electron distribution (in momentum states) as a whole cannot change.

If an electron with wave vector k is accelerated by the field to move into an already occupied state of k', then the electron at k' must hop to k and the total electron momentum in the valence band remains zero. The solid is an insulator. If a *high enough* field were applied, electrons might perhaps be ripped out of the valence band into the conduction band – a process known as dielectric breakdown. With the help of a band model we can see how the photoconductive effect works. The radiation is absorbed by electrons in the valence band and they are thus excited into the conduction band. The radiation changes the number of carriers available for conduction. Strictly Fig. 5.6(a) is the situation at absolute zero. However, the gap between the energy levels is so large that, even at higher temperatures, very few electrons are excited across the gap by thermal effects alone.

Figure 5.6(b) shows the energy bands of a semiconductor at absolute zero. This diagram looks very much like (a) except that the band gap is smaller. It is only the size of the gap which distinguishes between the two types of solid. A semiconductor is an insulator at absolute zero. As the temperature is increased an appreciable number of electrons may be excited across the gap. These electrons in the conduction band, together with the holes left behind in the valence band, can carry an electric current. The conductivity of a semiconductor is half way between that of a metal and that of an insulator. Also the conductivity of a semiconductor increases with temperature. The mobility of the charge carriers decreases with temperature, just as in the case of a metal, but the increase of the number of carriers available outweighs this.

Figure 5.6(c) shows the half-filled conduction band of an alkali metal. The electrons in the conduction band can be excited to the empty states in this band and so when a field is applied a current can flow. The charge is carried by electrons and, as there are many available, the resistivity of metals is low.

Semiconductors

We know that a semiconductor is a solid with energy bands *similar* to those represented schematically in Fig. 5.6(b). The resistivity is intermediate between that of insulators (10^{11} ohm-metres and more) and that of metals ($\sim 10^{-8}$ ohm-metres) at room temperature and it decreases with temperature. Nowadays semiconductors are much used in the manufacture of sophisticated electronic equipment. The electrical properties of the semiconductors used must be very accurately controlled.

Some semiconductors are pure chemical substances such as silicon and germanium (elements of group IV of the periodic table), GaAs (gallium arsenide, a III-V compound), and PbS (lead sulphide, a II-VI compound). These are known as *intrinsic* semiconductors.

It is now possible to grow pure single crystals of silicon and germanium with less than one impurity atom per 10^{10} atoms of solid. The intrinsic properties can be modified by the addition of known impurities in very carefully controlled amounts. These 'doped' crystals are known as *extrinsic* semiconductors.

Intrinsic semiconductors

The conductivity of the semiconductor will obviously depend on the number of charge carriers available. How can we calculate the number of electrons in the conduction band and the number of holes in the valence band?

A very simple model which is really a model of an insulator is shown in Fig. 5.7. The valence and conduction bands are assumed to be very narrow and relatively far apart. If the energy of *all* the states in the conduction band can be written as E_c and if there are Z allowed states per unit volume, then the number of electrons in the conduction band is given by

$$N_c = \frac{Z}{\exp\{(E_c - E_F)/k_B T\} + 1} \quad \text{(per unit volume).}$$

Similarly, the number of electrons per unit volume in the valence band is given by

$$N_v = \frac{Z}{\exp\{(E_v - E_F)/k_B T\} + 1};$$

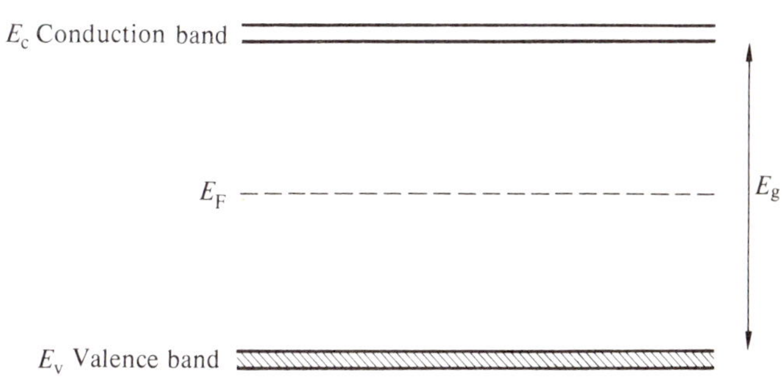

Fig. 5.7. Energy bands of an insulator.

where E_v is the energy of the valence band and the number of allowed states per unit volume is taken to be the same as for the conduction band. But what is E_F?

If we ignore the presence of any other valence or conduction bands then, because the number of electrons does not change,

$$N_c + N_v \equiv Z.$$

Then

$$\frac{1}{\exp\{(E_c - E_F)/k_BT\} + 1} + \frac{1}{\exp\{(E_v - E_F)/k_BT\} + 1} = 1$$

This can be rearranged to give

$$\exp\left\{\frac{E_c - E_F}{k_BT} + \frac{E_v - E_F}{k_BT}\right\} = 1$$

i.e.,

$$E_c - E_F + E_v - E_F = 0$$

or

$$E_F = \tfrac{1}{2}(E_v + E_c).$$

The Fermi level is halfway between the valence and the conduction bands. This is a result which you might have guessed before reading through the calculation. At absolute zero the valence band is completely filled and the conduction band is completely empty. The Fermi level must therefore be above the valence band but below the conduction band and it would not be expected to vary appreciably with temperature. In fact on this simple model E_F is temperature independent.

If $E_v - E_c$ is written as E_g, the energy of the band gap, then

$$N_c = \frac{Z}{\exp\{(E_c - E_F)/k_BT\} + 1} = \frac{Z}{\exp(E_g/2k_BT) + 1}$$

and if E_g is large compared with k_BT

$$N_c \sim Z \exp(-E_g/2k_BT).$$

The exponential dependence of N_c upon the temperature is, again, not very surprising. The conductivity σ is given by

$$\sigma = N_c e\mu_e + N_h e\mu_h$$

where N_h is the number of holes in the valence band ($=N_c$) and μ_e and μ_h are the respective mobilities of electrons in the conduction band and holes in the valence band. If we say that the mobilities are independent of temperature, this model predicts that a plot of $\log_e \sigma$ against $1/T$ should be a straight line. Such behaviour is observed experimentally. The reason why this simple model gives the right answer is that it makes two mistakes which cancel one another. The mobility is not independent of temperature – it varies approximately as $T^{-3/2}$ – and a better calculation of the carrier density shows a temperature dependence not only through the exponential term but also through a factor of $T^{+3/2}$. The more sophisticated calculation follows.

Although this calculation is better than the one on the insulator model it is still an approximation. The valence and conduction bands are assumed to be spherically symmetric, i.e., E is assumed to depend on $|\boldsymbol{k}|$ as measured from the maximum or minimum of the band. Then $Z(E)$, the number of states per unit volume per unit energy range, is taken to be equal to $CE^{1/2}$ where $C = (1/2\pi^2)(2m^*/\hbar^2)^{3/2}$, the same as the free electron case except that the electron mass is replaced by an effective mass m^*. The energies of the electrons in the conduction band are measured upwards from the bottom of the band; the energies of the holes in the valence band are measured downwards from the top of the band. We know that at absolute zero the Fermi level is somewhere between the two bands. The bands and energy levels are sketched in Fig. 5.8.

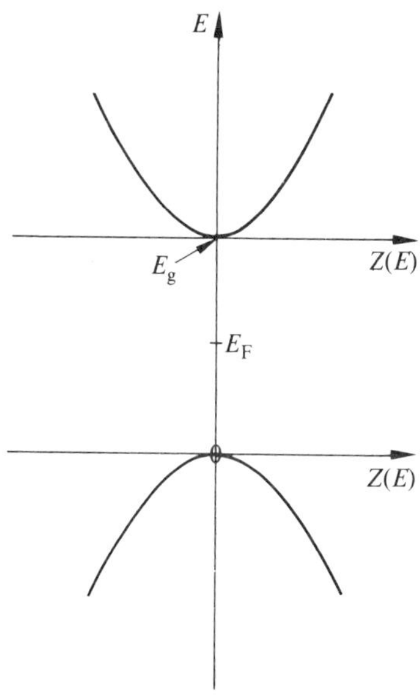

Fig. 5.8. Energy bands of a semiconductor.

If the zero of energy is taken to be at the top of the valence band, then N_e, the number of electrons per unit volume in the conduction band, is

$$N_e = \int_{E_g}^{\text{top of conduction band}} C_e(E - E_g)^{1/2} F(E)\, dE$$

where

$$C_e = \frac{1}{2\pi^2}\left(\frac{2m_e^*}{\hbar^2}\right)^{3/2}.$$

However, by taking the density of states to be spherically symmetric, we are in effect saying that only electrons near the bottom of the conduction band or holes near the top of the valence band are important. Under such circumstances the upper limit of the integration can be taken as infinity without significantly affecting the answer. Also, as in the last calculation, $E_g - E_F$ is taken to be sufficiently large to make the unity in the denominator of $F(E)$ unimportant.

$$N_e = C_e \int_{E_g}^{\infty} (E - E_g)^{1/2} \exp\{-(E - E_F)/k_B T\}\, dE.$$

Writing $(E - E_g)/k_B T = x^2$

N_e becomes equal to

$$C_e \exp\{(E_F - E_g)/k_B T\} \int_0^\infty x(k_B T)^{1/2} \exp(-x^2) 2x k_B T\, dx$$

$$= C_e (k_B T)^{3/2} \exp\{(E_F - E_g)/k_B T\} \int_0^\infty 2x^2 \exp(-x^2)\, dx$$

$$= C_e (k_B T)^{3/2} \exp\{(E_F - E_g)/k_B T\} \int_0^\infty \exp(-x^2)\, dx,$$

and the integral is a gamma function which is equal to $\sqrt{\pi}/2$.†

The number of holes in the valence band is given by

$$N_h = C_h \int_{\text{bottom of valence band}}^{0} (-E)^{1/2} \{1 - F(E)\} dE$$

where

$$C_h = \frac{1}{2\pi^2}\left(\frac{2m_h^*}{\hbar^2}\right)^{3/2}.$$

† $\int_0^\infty \exp(-x^2)\, dx \int_0^\infty \exp(-y^2)\, dy = I^2 = \int_0^\infty \int_0^\infty \exp\{-(x^2 + y^2)\}\, dx\, dy$
$= \int_0^\infty \exp(-r^2) \times \tfrac{1}{4} \times 2\pi r dr = \pi/4.$

$\{1 - F(E)\}$ is the probability that an electron state is unoccupied, i.e., the probability that there is a hole present. When $E - E_F$ is sufficiently large and negative

$$\frac{1}{\exp\{(E - E_F)/k_BT\} + 1} \sim 1 - \exp\{(E - E_F)/k_BT\}$$

and $\{1 - F(E)\} \sim \exp\{(E - E_F)/k_BT\}$. By the same argument as before, the lower limit of the integral may be taken to be minus infinity.

$$N_h = \int_{-\infty}^{0} C_h(-E)^{1/2} \exp\{(E - E_F)/k_BT\} \, dE$$

and if $x^2 = -E/k_BT$

$$N_h = C_h \exp\{-E_F/k_BT\} \int_{-\infty}^{0} x(k_BT)^{1/2} \exp(-x^2)\,(-2x)\,k_BT\,dx$$

$$= C_h(k_BT)^{3/2} \exp(-E_F/k_BT)\,\frac{\sqrt{\pi}}{2}.$$

If we multiply N_e by N_h we obtain the expression

$$N_eN_h = C_eC_h(k_BT)^3\,\frac{\pi}{4}\,\exp(-E_g/k_BT). \tag{5.7}$$

This expression does not depend on E_F. Nor have we made any assumptions about the relationship between N_e and N_h. Equation (5.7) applies not only in intrinsic semiconductors but also in extrinsic ones which we will come to later.

We know that in *intrinsic* semiconductors all the electrons in the conduction band have come from the valence band. Therefore $N_e = N_h$:

$$\frac{1}{2\pi^2}\left(\frac{2m_e^*}{\hbar^2}\right)^{3/2} (k_BT)^{3/2} \exp\{(E_F - E_g)/k_BT\}\,\frac{\sqrt{\pi}}{2}$$

$$= \frac{1}{2\pi^2}\left(\frac{2m_h^*}{\hbar^2}\right)^{3/2} (k_BT)^{3/2} \exp(-E_F/k_BT)\,\frac{\sqrt{\pi}}{2}$$

$$m_e^{*3/2} \exp\{(E_F - E_g)/k_BT\} = m_h^{*3/2} \exp(-E_F/k_BT)$$

$$\frac{3}{2}\log_e m_e^* + \frac{E_F - E_g}{k_BT} = \frac{3}{2}\log_e m_h^* - \frac{E_F}{k_BT}$$

or

$$E_F = \frac{E_g}{2} + \frac{3}{4}k_BT\log_e\frac{m_h^*}{m_e^*}.$$

When $T = 0$ K or when $m_h^* = m_e^*$, then E_F is halfway up the forbidden energy gap. Usually m_h^* is greater than m_e^* so that as the temperature rises E_F increases slowly.

Extrinsic semiconductors

We now have an idea of how the carrier density of intrinsic semiconductors should vary with temperature. In such solids the electrical properties depend upon the behaviour of the pure chemical compound. Extrinsic semiconduction is when the electrical properties depend strongly upon an impurity which is present in the lattice. Such impurities are introduced in very carefully controlled amounts during the manufacture of semiconductors for the electronics industry. (By impurity I mean a chemical impurity. Defects can cause extrinsic behaviour in semiconductors but this need not concern us.)

Let us consider silicon and germanium – two elemental semiconductors. Both are in Group IV of the periodic table. The crystals of these elements have the diamond structure (see chapter 1) in which each atom is surrounded by four nearest-neighbour atoms at the corners of a tetrahedron. A two-dimensional representation of a pure crystal is shown in Fig. 5.9(a). If a very small amount of a Group V element such as phosphorus, arsenic, or antimony is introduced into the crystal the lattice will not be significantly affected. The Group V atom will take its place in the lattice just as would another Group IV atom. But it has an extra electron above the number it needs to participate in the crystal bonds. (Fig. 5.9(b).) This extra electron is weakly bound to the impurity atom, for to leave it would mean creating a positive charge in the lattice. If the impurity atoms are sufficiently far apart, these extra, bound electrons form a set of localized states with energy levels just below the conduction band of the host semiconductor (Fig. 5.9(c)). The Group V atoms are known as donor atoms, the new energy levels as donor levels, and the doped semiconductor is referred to as n-type.

Figure 5.9(d) shows what happens when a Group III element such as boron, aluminium, indium, or gallium is added to the lattice. The foreign atoms have one electron less than they need to complete the crystal bonds and so, effectively, there is a hole or a vacant electron state centred on each atom. The new acceptor levels are shown in Fig. 5.9(e). The Group III atoms are known as acceptor atoms and the doped semiconductor is said to be p-type.

Under certain conditions we can calculate the carrier density and the Fermi levels in doped semiconductors. Suppose we have an n-type semiconductor for which the band gap between the valence and

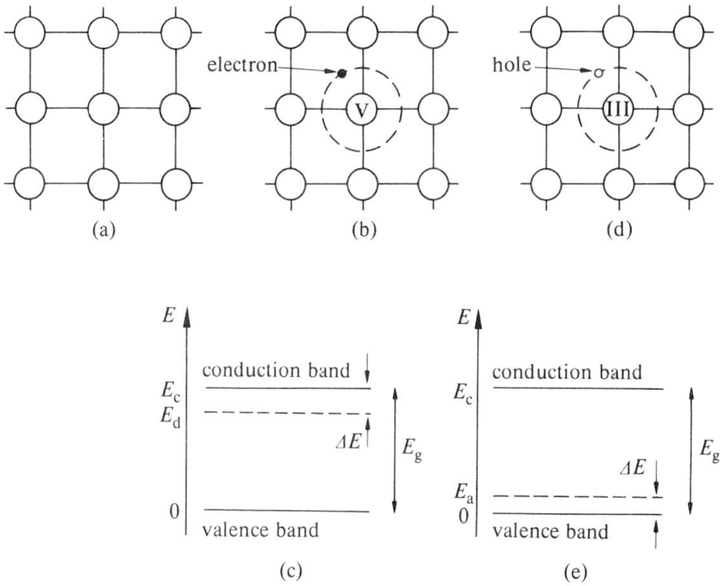

Fig. 5.9. Impurities in semiconductors. Open circles represent host atoms.

conduction bands is large compared with ΔE, the separation between the conduction band and the donor levels. We might therefore expect that, at low temperatures anyway, practically all the electrons in the conduction band would have come from the donor levels rather than from the valence band.

If there are N_d donor atoms per unit volume, then the number of ionized donors is just

$$N_d^+ = \{1 - F(E)\} N_d$$

$$= \left[1 - \frac{1}{\exp\{(E_d - E_F)/k_B T\} + 1}\right] N_d .$$

If the exponential term is sufficiently small

$$N_d^+ = N_d \exp\{(E_d - E_F)/k_B T\}.$$

The number of electrons per unit volume in the conduction band has already been calculated.

$$N_e = \frac{1}{2\pi^2}\left(\frac{2m_e^*}{\hbar^2}\right)^{3/2} (k_B T)^{3/2} \frac{\sqrt{\pi}}{2} \exp\{(E_F - E_g)/k_B T\}.$$

Writing $N_d^+ = N_e$ and rearranging gives the expression for the Fermi level:

$$E_F = \frac{E_d + E_g}{2} + \frac{k_B T}{2} \log_e \left\{ \frac{4N_d}{(2m_e^* k_B T/\hbar^2 \pi)^{3/2}} \right\}.$$

At $T = 0$ K, E_F is half way between the donor levels and the conduction band. Note that we have assumed that $\exp\{(E_d - E_F)/k_B T\}$ is small and therefore ΔE must be of the order of a few $k_B T$ for the calculation to be valid. At a finite temperature we can find the number of carriers by taking the square root of the product of the expressions for N_d^+ and N_e.

$$N_e = N_d^+ = \left[\frac{1}{2\pi^2} \left(\frac{2m_e^*}{\hbar^2} \right)^{3/2} (k_B T)^{3/2} \frac{\sqrt{\pi}}{2} N_d \exp\{(E_F - E_g)/k_B T\} \exp\{(E_d - E_F)/k_B T\} \right]^{1/2}$$

$$= \frac{N_d^{1/2}}{2} \left\{ \frac{2m_e^* k_B T}{\pi \hbar^2} \right\}^{3/4} \exp\{(-\Delta E)/2k_B T\}$$

where $\Delta E = E_g - E_d$.

Similar calculations on a p-type semiconductor show that at absolute zero the Fermi level is halfway between the valence band and the acceptor levels. Also, a similar expression holds for the number of holes in the valence band except that m_h^* replaces m_e^*, N_a replaces N_d, and ΔE is now the separation between the top of the valence band and the acceptor levels.

Variation of Fermi level with temperature. Figure 5.10(a) shows the electron distribution in an n-type semiconductor at absolute zero. The valence band and the donor levels are completely occupied and the conduction band is completely empty. E_F $(T = 0)$ is halfway across the gap between the donor levels and the conduction band. There are only a finite number of donor electrons and when the temperature is high enough all these electrons will have been excited into the conduction band. Further electrons arriving in the conduction band must be from the valence band and gradually, as T increases, the intrinsic properties will become more and more important. The Fermi level will eventually fall towards E_F (intrinsic) which itself varies slightly with temperature. Figure 5.10(b) shows the equivalent diagram for a p-type semiconductor. In the diagrams the fact that E_g probably decreases with temperature has been ignored.

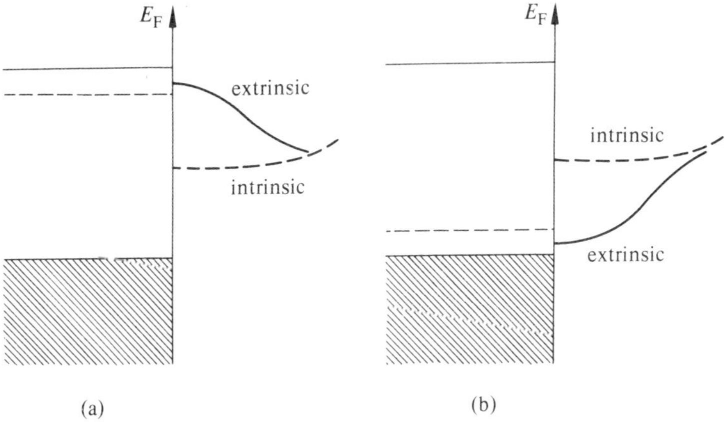

Fig. 5.10. Variation of Fermi level with temperature for (a) an n-type, (b) a p-type semiconductor.

Measurement of semiconductor properties

The effective mass. The effective masses of electrons and holes in semiconductors are measured by cyclotron resonance. If a magnetic field is applied at right angles to the direction of motion of a charged particle, the particle will describe a circular path with angular frequency

$$\omega = \frac{eB}{m^*}.$$

If radiation of the same angular frequency is supplied, resonant absorption can take place. Usually a fixed frequency (of about 10^{10} Hz) is used and the magnetic field is swept (up to about 0.5 tesla) while absorption occurs. The mean free path of the carriers should be great enough for them to complete at least one revolution before collision. This means working at liquid helium temperatures where the density of carriers is extremely small. The number of carriers can be increased by irradiating the sample with light of a sufficiently high frequency to cause excitation across the band gap.

If the bands were spherically symmetric near single edges, as has been assumed in the calculations of previous sections, then one would obtain only two absorption peaks from cyclotron resonance experiments. One of these would correspond to electrons of m_e^* and the other to holes of m_h^*. Unfortunately things are not so simple in real crystals.

Let us consider what happens in the case of silicon and germanium. Around a minimum in the conduction band the energies of the carriers can be expressed as

$$E = \frac{\hbar^2}{2}\left(\frac{k_x^{\,2}}{m_x{}^*} + \frac{k_y^{\,2}}{m_y{}^*} + \frac{k_z^{\,2}}{m_z{}^*}\right)$$

provided that the axes are chosen properly. Then, with B applied along the z-direction, $\omega = eB/(m_x{}^* m_y{}^*)^{1/2}$. Similar equations apply in the other two directions. Three values of $m_e{}^*$ can be obtained. As it happens, for silicon and germanium, two of the $m_e{}^*$s are the same. The situation is more complicated in the valence band because two bands of different curvature meet at the band edge. We then get two separate lots of holes – 'heavy' holes and 'light' holes – from the different bands.

The effective mass of a carrier is obviously a tensor rather than a scalar quantity. Our calculations of carrier density were oversimplified. However, N is a scalar and one would expect the formulae still to hold provided that the right average values for $m_e{}^*$ and $m_h{}^*$ are used. A much worse problem is what to do about calculating conductivity. Conductivity depends on the mobility which in turn depends on m^* ($\mu = |e|\tau/m^*$) and it therefore might appear that the mobility and the conductivity should be anisotropic. When conductivity is measured it is not found to be anisotropic. What is going on? Well, it is perfectly true that the mobility of the electrons at a minimum in the conduction band is a tensor and very anisotropic. But (see Fig. 5.11) the minimum in the conduction band does *not* occur for $\boldsymbol{k} = 0$. For germanium (with two atoms per unit cell) there are eight minima along the 1 1 1 directions in the Brillouin zone; for silicon there are six minima along the 100 directions. The x-, y-, and z-directions for one minimum are not the same as for another. When we average over all the minima we find that the mobility is isotropic and depends on an average $m_e{}^*$ given by

$$\frac{1}{m_e{}^*} = \frac{1}{3}\left(\frac{1}{m_x{}^*} + \frac{1}{m_y{}^*} + \frac{1}{m_z{}^*}\right).$$

We can happily go ahead using a scalar average $m_e{}^*$ and a scalar mobility.

Table 5.1 shows $m_x{}^*, m_y{}^*, m_z{}^*$, and an average value $m_e{}^*$; also shown are average masses of the light and heavy holes (the surfaces here are not quite spheres). Note that the values given are all less than m, the ordinary electron mass.

The absorption curves (with B in different crystal directions) provide information on the shape of the actual E versus $\boldsymbol{k}$ curves. The

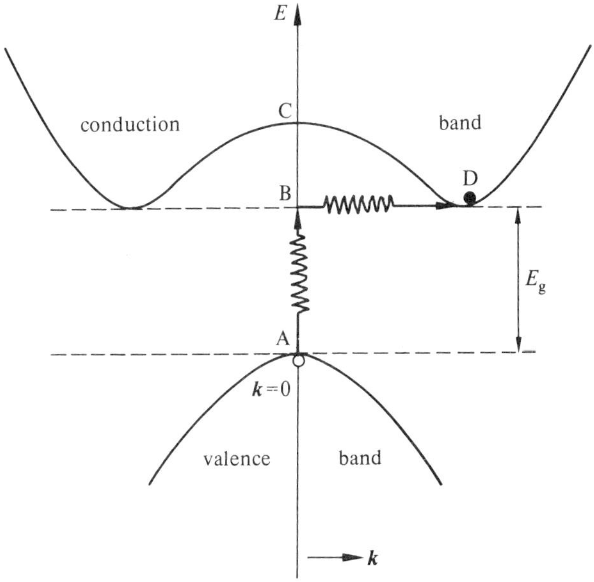

Fig. 5.11. Schematic representation of energy bands in indirect-gap semiconductor such as silicon or germanium.

area of the peaks indicates the number of carriers involved, and the width their lifetime before collision.

TABLE 5.1

	Electron mass/m		Heavy hole mass/m	Light hole mass/m	Band gap at room temp.
Si	0.98, 0.19, 0.19	0.26	0.50	0.16	~1.1 eV
Ge	1.64, 0.08, 0.08	0.12	0.28	0.04	~0.7 eV
GaAs		0.07	0.65	0.07	~1.4 eV

The band gap. One way to arrive at an estimate of E_g is to study the curves of $\log_e(\sigma)$ against $1/T$ in the intrinsic region. Another way is to measure the threshold of optical absorption of the solids. They will begin to absorb radiation when the photons have sufficient energy to excite electrons across the band gap. Some semiconductors, such as GaAs, are called *direct-gap* semiconductors. This is because the minimum of the conduction band and the maximum of the valence band occur for the same value of $\boldsymbol{k}$. Because a photon of band gap energy has very little momentum, transitions are 'vertical' on an E versus k diagram. The threshold energy corresponds to E_g. Silicon and germanium are examples of *indirect-gap* semiconductors. The threshold energy does not correspond to AC (Fig. 5.11) although strong absorption does begin to take place at those energies. Absorption begins at an energy very close to E_g. Phonons are involved in the absorption process. A photon can take the electron to B and a phonon provides the necessary momentum to take the electron to D where there is an allowed state available. The phonon energy is small so BD is shown 'horizontal' on the diagram.

At temperatures high enough for the necessary phonons to be readily available, a phonon may be absorbed as shown.

Threshold energy $= E_g -$ energy of phonon with wavevector corresponding to BD.

At low temperatures the phonon will have to be emitted as there will be few available for absorption and the threshold energy will be given by

Threshold energy $= E_g +$ energy of phonon with wavevector corresponding to BD.

The mobility. The mobility can be measured for minority carriers in a way which is represented in Fig. 5.12. A slab of n-type semiconductor is connected between two terminals, C and D, so that a current flows in the direction indicated. At intervals of approximately a microsecond a positive pulse is applied at the terminal A. This pulse is sufficiently positive to draw electrons from the valence band of the semiconductor as well as from the conduction band. When the potential pulse is applied it travels instantaneously along to B (actually with the speed of light) and all the conduction electrons in the n-type semiconductor begin to move simultaneously. The number of electrons that are drawn

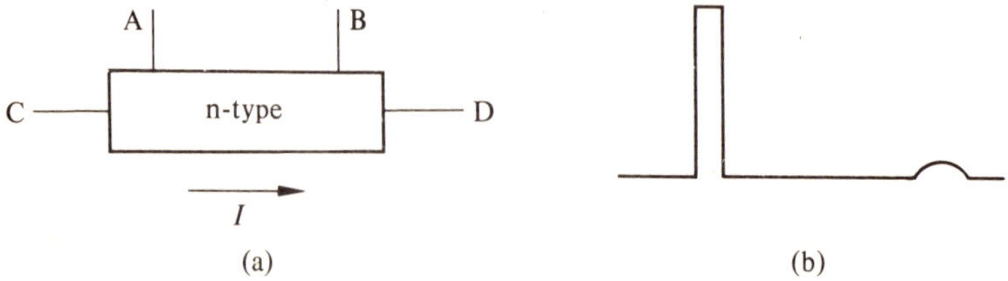

Fig. 5.12. (a) Sketch of apparatus for measuring mobility. (b) Pulse on oscilloscope screen.

from B depends upon the size of the pulse at A and on the mobility of the electrons in the conduction band. Because the electrons enter into the conduction band at B at the same time as those leave at A, we cannot measure the mobility of the electrons.

The holes in the valence band at A begin to drift under the potential across CD towards B. As there are normally no appreciable number of holes in an n-type semiconductor there will be no hole current at B until the holes at A have drifted to B. If there were holes present throughout the sample then the effect would be undetectable. (A local excess of carriers is dissipated in a time which depends on the inverse of the number of similar carriers present.) At a time equal to the length AB divided by the hole drift velocity a second pulse reaches B. The size of the pulse will depend on AB for the holes are not in thermal equilibrium with the lattice and will gradually combine with electrons from the valence band and be annihilated. The two pulses – one due to electrons in the conduction band and the other to holes in the valence band – can be displayed on an oscilloscope screen.

The mobility is calculated from the fact that it is the drift velocity in unit field or the observed drift velocity divided by the field across AB. The mobility of electrons in the conduction band of a p-type semiconductor (again the minority carriers) can be measured by a similar technique.

As an example of the magnitudes involved, the mobility of electrons in silicon at room temperature is $\sim 13 \times 10^{-2}$ m^2 V^{-1} s^{-1}, and of holes $\sim 5 \times 10^{-2}$ m^2 V^{-1} s^{-1}. For germanium the corresponding values are $\sim 36 \times 10^{-2}$ m^2 V^{-1} s^{-1} and $\sim 18 \times 10^{-2}$ m^2 V^{-1} s^{-1}. Note that the electrons are more mobile than the holes. This is what one would expect knowing that m_e^* is less than m_h^*. The mobility of electrons in a metal is much smaller ($\sim 0.3 \times 10^{-2}$ m^2 V^{-1} s^{-1} for copper). Metals are good conductors not because of the mobility of their charge carriers but because there are so many carriers available.

The carrier density. We can measure the mobility μ_e of electrons in a p-type semiconductor and the mobility μ_h of holes in an n-type. For an intrinsic semiconductor $\sigma = Ne\,(\mu_e' + \mu_h')$ where N is equal to the number of electrons in the conduction band and to the number of holes in the valence band. μ_e and μ_h have been measured in the presence of impurities and may possibly differ from μ_e' and μ_h'. If this difference is ignored, N may be calculated from the conductivity.

When the semiconductor is highly doped so that only one type of carrier is present, it is fairly easy to find N direct from the Hall coefficient.

Conductivity. The conductivity of a semiconductor, measured at various temperatures, is typically of the form shown in Fig. 5.13. At low temperatures extrinsic semiconduction will be dominant, for even in 'pure' samples there are some defects or impurities. The curve $\log_e(\sigma)$ against $1/T$ has a slope of $-\Delta E/2$ where ΔE is the separation between donor levels and conduction band or between valence band and acceptor levels. At slightly higher temperatures all the impurities will be ionized (donors) or in possession of electrons (acceptors), but there will be very few carriers excited across the band gap. In this region the number of charge carriers is practically independent of temperature and the conductivity drops because of the reduction of mobility as T increases. At still higher temperatures intrinsic behaviour is dominant and $\log_e(\sigma)$ against $1/T$ has a steeper slope of $-E_g/2$.

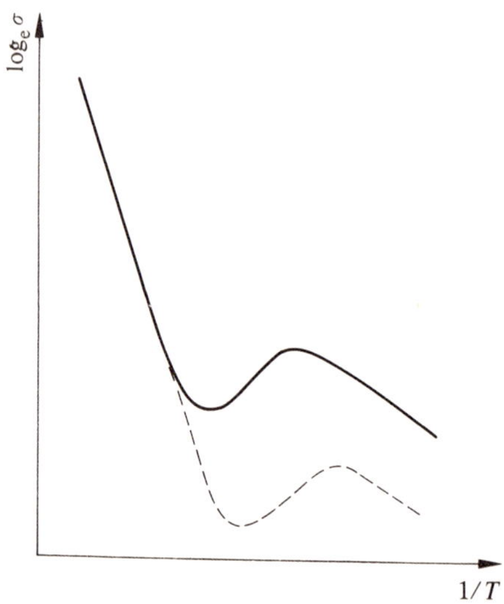

Fig. 5.13. The temperature dependence of the conductivity of a semiconductor.

The p–n junction

The p–n junction rectifies. The vital steps in the understanding of this process are sketched in Fig. 5.14. (a) shows the energy levels in a slab of p-type semiconductor and in a slab of n-type semiconductor. (b) shows what happens when these two slabs are in contact. (Physical contact is not enough – the slabs must be grown together.) A similar process takes place to that described in chapter 4 when metals with different work functions are in contact with one another. Electrons flow from the

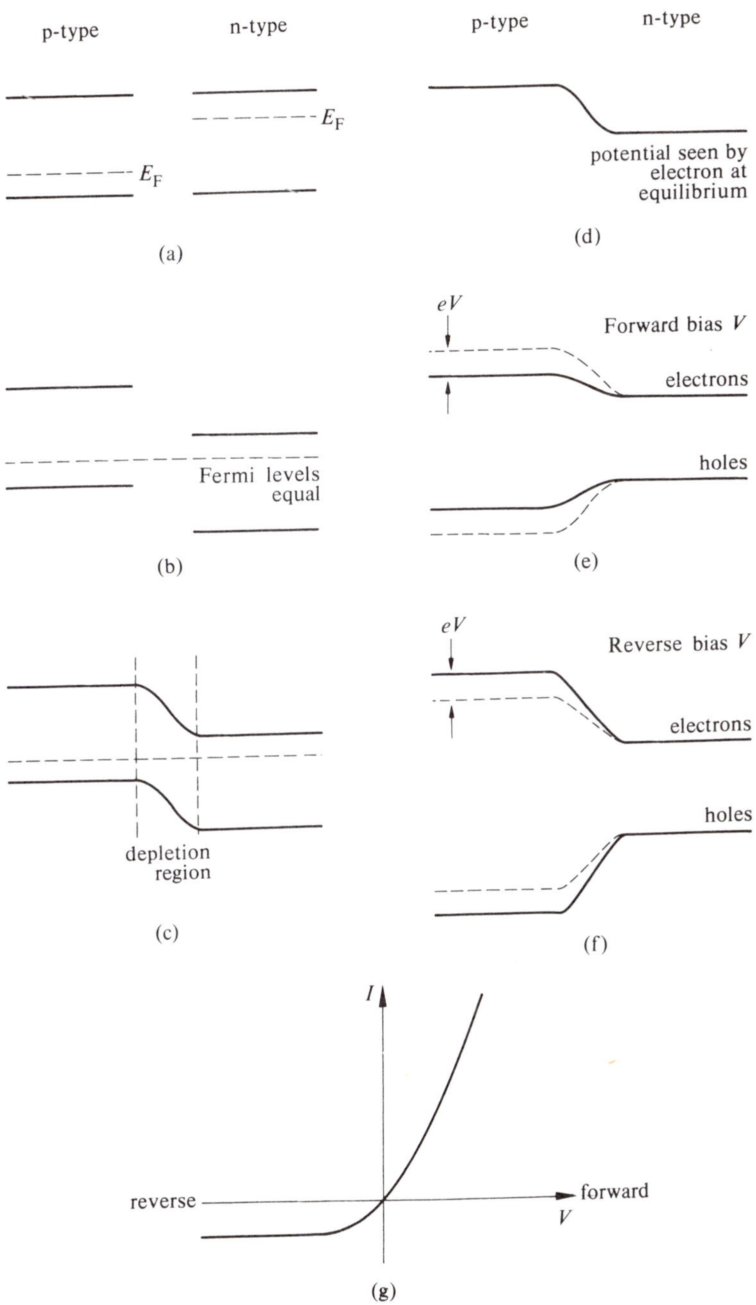

Fig. 5.14. The p–n junction.

n-type to the p-type semiconductor until the Fermi levels are equalized. The result of this is that the p-type slab gains a negative charge near its surface whereas the n-type slab obtains a positive charge. Unlike the situation which occurs in metals, this charge cannot reside *at* the surface but penetrates into the two slabs through a small distance at each side of the boundary. In this region the bands are bent. This phenomenon is shown in (c). The electrons in the conduction band of the n-type are repelled from the region and there is an increase in the number of holes in the valence band. However, because $N_e N_h$ is constant, the total number of charge carriers is reduced. A similar situation occurs near the boundary in the p-type slab. This region is therefore known as the depletion region.

At equilibrium, the number of electrons flowing across the junction from the n-type material to the p-type is small because there is a potential barrier for them to overcome (d). It is exactly equal to the small number of electrons which cross in the opposite direction. Only a few electrons are in the conduction band of the p-type material but all those that move towards the junction will cross by 'running down the potential hill'. Also the number of holes crossing from right to left is equal to the number crossing from left to right. The energy levels shown in (a) to (c) are those for electrons, i.e., minus the energies for holes.

Suppose now that a forward-bias voltage is connected across the device. The p-type material is made positive with respect to the n-type. The voltage drop will be mainly across the depletion region because of the lack of carriers there. (e) shows the new potential as seen by electrons and by holes. Any holes in the n-type material will still be able to get across to the p-type side and the same applies to the electrons in the conduction band of the p-type material. They will also be able to cross the junction. But these are minority carriers. The many electrons in the n-type slab will find that the potential barrier preventing them crossing the junction has been lowered and the holes in the p-type are in the same situation. Because the number of carriers is roughly exponentially dependent on the distance from the band edges, there will be an exponential increase in the current due to the majority carriers (the main part of the current).

Diagram (f) shows the situation when reverse-bias voltage is applied and the p-type side is made negative. Again, the current due to the minority carriers is not much changed but there is an exponential decrease of the current carried by the majority carriers. As the reverse bias is increased a small saturation current is reached. This is due entirely to minority carriers.

(g) shows a plot of voltage against current. The junction is non-ohmic for the resistance depends on the direction of the current flow. The junction is said to rectify. It is an example of the type of electronic device that can be manufactured from doped semiconductor crystals.

Examples

1) 1969 Paper IVA

A sample of silicon contains 10^{-4} atomic per cent of phosphorus donors which are all singly ionized at room temperature. The band gap of silicon is 1.1 eV and the electron mobility is 1500 $cm^3\ V^{-1}\ s^{-1}$. Calculate the resistivity of the sample.

(For Si: atomic weight = 28, density = 2.3 g cm^{-3}.)

Answer

$$\text{Resistivity} = \frac{1}{ne\,\mu}$$

where e is the charge on the electron = 1.6×10^{-19} C, μ is the mobility = $1500 \times 10^{-4}\ m^2\ V^{-1}\ s^{-1}$, and n is the number of carriers per unit volume. (At room temperature, and with a band gap of 1.1 eV, effectively *all* the carriers will come from the donors.)

$$n = \text{number of electrons per } m^3 = \frac{10^{-4}}{100} \times \text{number of Si atoms per } m^3.$$

$$= 10^{-6} \times \frac{2.3 \times 10^6\ g\ m^{-3}.}{28 \times 1.66 \times 10^{-24}\ g.}$$

$$\text{Resistivity} = \frac{28 \times 1.66 \times 10^{-24}}{2.3} \times \frac{1}{1.6 \times 10^{-19}} \times \frac{1}{0.15}\ m^3 C^{-1} m^{-2} V$$

$$= 8 \times 10^{-4}\ \frac{m\,V}{C\,s^{-1}}$$

$$= 8 \times 10^{-4}\ \text{ohm m}.$$

2) 1970 Pure Physics Option I

In a cyclotron resonance experiment on silicon at 2.4×10^{10} Hz a resonance is found at 4300 gauss (0.43T). Calculate the value of m^* for the charge carrier responsible for this resonance in units of the free electron mass, m_e.

Answer

When a magnetic field is applied at right angles to the direction of motion of a charged particle, the particle will describe a circular path with angular velocity

$$\omega = \frac{eB}{m^*}.$$

Of course, this strictly applies only to free particles and not to those in solids but it should be a reasonable approximation.

$$2\pi \times 2.4 \times 10^{10} = \frac{1.60 \times 10^{-19} \times 0.43}{m^*}$$

$$\frac{m^*}{m_e} = \frac{1.60 \times 10^{-19} \times 0.43}{2\pi \times 2.4 \times 10^{10} \times 9.11 \times 10^{-31}} = 0.5$$

Questions

1) *Manchester, 1971*

What are the Brillouin zones? Why are they a useful concept in solid state physics?

Calculate the ratio k_F/k_m for a monovalent metal with the face-centred-cubic structure, where k_F is the free-electron Fermi wave-vector k_m is the minimum distance in k-space between the surface of the first-Brillouin zone and the origin of k-space.

Would you expect your calculations to apply to the Fermi surface of a real metal such as copper?

2) *Manchester, 1971*

Three samples of germanium doped with arsenic have donor concentrations N_D of 5.5 10^{16} cm^{-3}, 1.7 10^{15} cm^{-3} and 1.4 10^{14} cm^{-3}. All samples have acceptor concentrations N_A less than 10^{13} cm^{-3}.

Sketch the behaviour of n, the number of mobile carriers cm^{-3}, for the three samples as a function of T^{-1} where T is the absolute temperature. Use a logarithmic scale for the n-axis. Give physical reasons for the main features of your sketch.

Use the following data for germanium:
valence band/conduction band energy gap E_G = 0.67 eV,
donor ionization energy E_D for As impurities = 0.0127 eV,
intrinsic electron density n_i at room temperature = 2.5 10^{13} cm^{-3}.

3) *Manchester, 1971*

How can information about the Fermi surfaces of metals be gained from cyclotron resonance experiments? Describe the experimental details and conditions necessary to carry out such experiments.

In such an experiment on potassium at 68 10^9 Hz, three consecutive resonances were observed at magnetic fields of 0.73, 0.59, and 0.49

tesla (weber m^{-2}). Calculate the cyclotron effective mass of electrons in potassium.

[Hint: the general condition for resonance is $B = \dfrac{\omega m}{e n}$ where n is an integer.]

4) *Manchester, 1974*
A piece of n-type silicon has an arsenic concentration of 10^{16} cm^{-3} and a boron concentration of $4\ 10^{15}$ cm^{-3}. The intrinsic density of carriers is $3\ 10^{10}$ cm^{-3} at room temperature. Estimate the density of ionized donors, ionized acceptors, majority carriers, and minority carriers at room temperature.

5) *Manchester, 1971*
Draw the reciprocal lattice and first two Brillouin zones in the extended zone scheme for a two-dimensional square lattice of divalent atoms, with lattice constant a = 2.5 Å.

Indicate the regions of k-space corresponding to occupied electron states in the cases of (a) an insulator and (b) a metal, commenting on the influence of the periodic lattice potential on the energies of the electron states.

6) *Manchester, 1972*
Explain the terms Bravais lattice, reciprocal lattice, Brillouin zone. The atoms of a fictitious two-dimensional monovalent metal are arranged in a square lattice. Sketch a few reciprocal lattice sites, showing the boundaries of the first Brillouin zone. Calculate the free-electron Fermi wave-vector and hence shade in the region of k-space occupied by electrons in this zone.

This density of the metal is increased so that extra atoms of the same metal occupy the centre of each of the cells in the original square lattice. What is the Bravais lattice of the new structure? Does the Fermi surface overlap into the second Brillouin zone? Give reasons for your answer.

Reference

1. B. I. Bleany and B. Bleany. *Electricity and Magnetism.* Clarendon Press, 1965.

6. Magnetic Properties of Solids

Diamagnetism

All atoms, whether or not they are paramagnetic, make a small negative diamagnetic contribution to the magnetic susceptibility. All electron orbits are affected by the presence of an applied magnetic field. The planes of the orbits precess and this gives rise to an effective current loop. The direction of the current is, by Lenz's law, such as to oppose the flux through the orbit. Thus, the contribution to the susceptibility is negative. I mention this only to remind you that experimental measurements of susceptibility of solids may have to be corrected for the diamagnetic term before they can be accurately compared with theory.

Paramagnetism

Classical theory. Paramagnetism is due to the existence of permanent magnetic dipoles associated with atoms, ions, or molecules. (See also chapter 4 for the paramagnetism of free electrons.) It is found experimentally that for many substances in fairly weak magnetic fields the susceptibility is inversely proportional to temperature. This behaviour is known as Curie's law. Langevin provided a theoretical interpretation of this behaviour. He made use of classical statistics and the Boltzmann distribution function.

The number of magnetic dipoles, m, making an angle θ with the magnetic field B will be given by

$$\mathrm{d}n = AZ(E)\exp(-E/k_{\mathrm{B}}T)\,\mathrm{d}E$$

where A is a constant, E is the potential energy of a dipole in the field, and $Z(E)\,\mathrm{d}E$ is equal to the number of energy states between E and $E + \mathrm{d}E$. Now $E = -mB\cos\theta$ (in SI units, B is the force vector) and so we can write

$$\mathrm{d}n = A'\,Z(\theta)\exp(mB\cos\theta/k_{\mathrm{B}}T)\,\mathrm{d}\theta$$

where now $Z(\theta)\,\mathrm{d}\theta$ is the number of states between θ and $\theta + \mathrm{d}\theta$. If the magnetic dipole can have any projection upon the field direction then (see Fig. 6.1) the number of energy states between θ and $\theta + \mathrm{d}\theta$ will be simply

$$\frac{\text{area of the sphere surface between } \theta \text{ and } \theta + \mathrm{d}\theta}{\text{total area of sphere}}$$

$$= \frac{2\pi\, m \sin\theta \cdot m\, \mathrm{d}\theta}{4\pi m^2} = \frac{\sin\theta\, \mathrm{d}\theta}{2}\,.$$

So

$$\mathrm{d}n = A'' \sin\theta \exp(mB\cos\theta/k_{\mathrm{B}}T)\,\mathrm{d}\theta\,.$$

Then the average projection on the field direction is

$$\overline{m} = \frac{1}{n}\int_0^\pi m\cos\theta\,\mathrm{d}n$$

where n is equal to the total number of dipoles and is therefore given by

$$n = \int_0^\pi \mathrm{d}n.$$

$$\frac{\overline{m}}{m} = \frac{\int_0^\pi \cos\theta\sin\theta\exp(mB\cos\theta/k_{\mathrm{B}}T)\,\mathrm{d}\theta}{\int_0^\pi \sin\theta\exp(mB\cos\theta/k_{\mathrm{B}}T)\,\mathrm{d}\theta}\,. \tag{6.1}$$

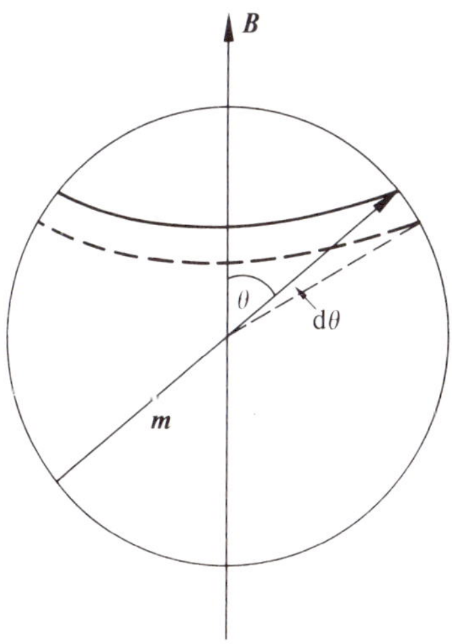

Fig. 6.1. The classical theory of paramagnetism. The dipole m is at an angle θ to the magnetic field B.

Substituting $y = mB/k_BT$ and $x = \cos\theta$, $-\sin\theta\ d\theta = dx$ and

$$\frac{\bar{m}}{m} = \frac{\int_{-1}^{1} x \exp(xy)\,dx}{\int_{-1}^{1} \exp(xy)\,dx}$$

$$= \frac{e^y + e^{-y}}{e^y - e^{-y}} - \frac{1}{y} = \coth y - \frac{1}{y} \quad \text{or} \quad L(y)$$

where $L(y)$ is known as the Langevin function. When mB is small in comparison with k_BT, y is small and $L(y)$ can be written as

$$\frac{2 + y^2 + \ldots}{2y + y^3/3 + \ldots} - \frac{1}{y} = \frac{1}{y}\left(1 + \frac{y^2}{2} + \ldots\right)\left(1 - \frac{y^2}{6}\right) - \frac{1}{y}$$

$$\approx \frac{y}{2} - \frac{y}{6} = \frac{y}{3}.$$

The susceptibility is defined as $N\bar{m}/H$ where N is the number of dipoles per unit volume. If the susceptibility is small (effectively we are ignoring any field acting on a dipole other than the applied field) then $H = B/\mu_0$ and

$$\chi = \text{susceptibility} = \frac{N\bar{m}\mu_0}{B} = Nm\frac{y}{3}\mu_0 \Big/ B = Nm\mu_0 \frac{mB}{3k_BT} \Big/ B$$

$$= \frac{Nm^2}{3k_BT}\mu_0$$

which is independent of the applied field and is inversely proportional to the absolute temperature. We can put $C = Nm^2\mu_0/3k_B$ and then $\chi = C/T$ which is Curie's law. C is known as the Curie constant. In the other extreme, when B is large and T small so that $y \to \infty$, $L(y) \to 1$. This means that $\bar{m}/m \to 1$, i.e., a saturation magnetization of Nm occurs where the dipoles are all lined up along the field direction.

Quantum theory. Unfortunately, the use of classical statistics is not always justified. We know from atomic physics and from quantum mechanics that the magnetic moment associated with a particular vector $\boldsymbol{J}$ (angular momentum $= \{J(J+1)\}^{1/2}\hbar$) cannot have just any projection upon the field direction. The magnetic moment can be written as $-g\beta\{J(J+1)\}^{1/2}$ where g is called the Landé 'g-factor' and β is a unit of magnetic moment known as the Bohr magneton. The projection of J upon B is equal to M_J where $M_J = +J, J-1, \ldots 0, \ldots -J$. Therefore, the integral of eq. (6.1) must be replaced by a summation over the allowed values of M_J. The energy E is now $M_J g\beta B = -\boldsymbol{m} \cdot \boldsymbol{B} = -mB\cos\theta$. We have

$$\overline{m} = \frac{\sum_{-J}^{+J} (-M_J g\beta) \exp(-M_J g\beta B/k_B T)}{\sum_{-J}^{+J} \exp(-M_J g\beta B/k_B T)} . \qquad (6.2)$$

Some manipulation is needed in order to tidy up this expression. With $y = Jg\beta B/k_B T$ it can be seen that the denominator is a geometric series of the form

$$\exp(y) + \exp\left(y - \frac{y}{J}\right) + \ldots \text{ with } 2J+1 \text{ terms.}$$

The sum of a geometric series of the form

$$a + ar + ar^2 + \ldots \text{ to } n \text{ terms is } a(1-r^n)/(1-r)$$

and therefore the denominator of eq. (6.2) is equal to

$$\frac{\exp(y)\left\{1 - \exp\left(-\frac{2J+1}{J}y\right)\right\}}{\left\{1 - \exp\left(-\frac{y}{J}\right)\right\}}.$$

Also, the numerator of eq. (6.2) will be seen, upon inspection, to be equal to $g\beta J$ times the differential of the denominator with respect to y. It is then equal to

$$g\beta J\left[\frac{\exp(y)\left\{1 - \exp\left(-\frac{2J+1}{J}y\right)\right\} + \exp(y)\cdot\frac{2J+1}{J}\cdot\exp\left(\frac{2J+1}{J}y\right)}{\left\{1 - \exp\left(-\frac{y}{J}\right)\right\}}\right]$$

$$+ g\beta J\left[\exp(y)\left\{1 - \exp\left(-\frac{2J+1}{J}y\right)\right\}\cdot\frac{-1}{\left\{1 - \exp\left(-\frac{y}{J}\right)\right\}^2}\cdot\frac{1}{J}\exp\left(-\frac{y}{J}\right)\right].$$

Equation (6.2) is

$$g\beta J\left[1+\frac{\frac{2J+1}{J}\exp\left(-\frac{2J+1}{J}y\right)}{\left\{1-\exp\left(-\frac{2J+1}{J}y\right)\right\}}-\frac{\frac{1}{J}\exp\left(-\frac{y}{J}\right)}{\left\{1-\exp\left(-\frac{y}{J}\right)\right\}}\right]$$

$$=g\beta J\left[\frac{2J+1}{2J}+\frac{2J+1}{2J}\cdot\frac{2\exp\left(-\frac{2J+1}{J}y\right)}{\left\{1-\exp\left(-\frac{2J+1}{J}y\right)\right\}}-\frac{1}{2J}-\frac{1}{2J}\cdot\frac{2\exp\left(-\frac{y}{J}\right)}{1-\exp\left(-\frac{y}{J}\right)}\right]$$

$$=g\beta J\left[\frac{2J+1}{2J}\cdot\coth\left(\frac{2J+1}{2J}y\right)-\frac{1}{2J}\cdot\coth\left(\frac{y}{2J}\right)\right]. \qquad (6.3)$$

The term multiplying $g\beta J$ is known as the Brillouin function – I shall refer to it as $Br(y)$ – and when J is very large it approaches the Langevin function $L(y)$, as one might expect. (When $J\to\infty$ the projections of the magnetic moment upon the field direction form a continuum.) $\chi = N\bar{m}/H$ which, under the same conditions as before $= N\bar{m}\mu_0/B$. When $Jg\beta B/k_BT$ is small (normal field strengths, 1 weber m^{-2}; ordinary temperatures, 300 K)

$$\coth\left(\frac{2J+1}{2J}y\right)\to\frac{2J}{(2J+1)y}+\frac{2J+1}{2J}\cdot\frac{y}{3}$$

$$\coth\left(\frac{y}{2J}\right)\to\frac{2J}{y}+\frac{y}{2J}\cdot\frac{1}{3}.$$

Therefore,

$$Br(y)\to\frac{1}{y}+\left(\frac{2J+1}{2J}\right)^2\frac{y}{3}-\frac{1}{y}-\left(\frac{1}{2J}\right)^2\frac{y}{3}=\frac{J(J+1)}{J^2}\frac{y}{3}.$$

Then

$$\chi=\frac{Ng\beta JBr(y)\mu_0}{B}=Ng\beta J\cdot\frac{(J+1)}{J}\cdot\frac{\mu_0}{B}\cdot\frac{Jg\beta B}{3k_BT}=\frac{Ng^2\beta^2J(J+1)}{3k_BT}\mu_0 \qquad (6.4)$$

which obeys Curie's law and is what one would expect from the classical theory with $m^2 = g^2\beta^2J(J+1)$. J need not be large to give this high-temperature correspondence.

As $T \rightarrow 0$ K and $y \rightarrow \infty$

$$\coth\left(\frac{2J+1}{2J}y\right) \rightarrow 1, \quad \coth\left(\frac{y}{2J}\right) \rightarrow 1,$$

and therefore

$$Br(y) \rightarrow \frac{2J+1}{2J} - \frac{1}{2J} = 1$$

and $N\bar{m} = Ng\beta J$. This does not correspond with the classical value of Nm, if m is put equal to $g\beta\{J(J+1)\}^{1/2}$, unless J is very large. This is because the maximum projection that the magnetic moment can have upon the field direction is $g\beta J$.

Paramagnetism in solids

Which solids show paramagnetic behaviour? We know from chapter 4 that the partly filled conduction band of metals is responsible for an electronic paramagnetism which is nearly independent of temperature. However, many insulating solids show a Curie-type behaviour with a paramagnetic susceptibility that decreases as temperature rises. When an atom is bound into an insulating solid it is usually the case that the electrons form closed-shell noble-gas structures or, to be more precise, that the energy bands corresponding to each sub-shell are completely filled. This situation results in *zero* net angular momentum and hence *zero* magnetic moment associated with the atoms. The only atoms† that possess a non-zero angular momentum are those with incomplete inner sub-shells such as the atoms of the transition series.

You may have got the impression from chapter 1 that the electron states, as we progress through the atoms of the periodic table, are filled up in the order $1s^2$, $2s^2$, $2p^6$, $3s^2$, $3p^6$, $3d^{10}$, $4s^2$, $4p^6$, $4d^{10}$, $4f^{14}$, ... ns^2, np^6, ... (where $1s^2$ indicates that two electrons can be placed in an s-state with principal quantum number 1, etc.). It may also have seemed that when each shell of a principal quantum number n was filled then we would have a noble gas. Things are not quite so simple. It is true that helium has an electronic configuration of $1s^2$ and that neon has $1s^2\ 2s^2\ 2p^6$ but argon has $1s^2\ 2s^2\ 2p^6\ 3s^2\ 3p^6$. The 3s and 3p sub-shells are complete but not the whole shell with $n = 3$. There is no stable, neutral, atom which has a configuration of $1s^2\ 2s^2\ 2p^6\ 3s^2\ 3p^6\ 3d^{10}$ although this electron pattern is shared by Cu^+ and Zn^{++}. The element following argon in the periodic table is potassium and this is not

†I use the word in its general sense to include ions in ionic compounds.

. . . $3s^2$ $3p^6$ $3d^1$ but . . . $3s^2$ $3p^6$ $4s^1$. Unlike the energy levels of the hydrogen atom, here the 4s orbit has lower energy than the 3d. We can explain this by saying that the 4s orbit which penetrates deeper into the atom core than does the 3d orbit, is less shielded from the attractive nuclear charge and therefore its energy is lowered. Calcium, with atomic number 20, follows potassium and has the electronic configuration . . . $3s^2$ $3p^6$ $4s^2$. At this stage the 3d orbits begin to fill. The nuclear charge has increased and their energy is lower than that of the 4p orbits. Scandium (21) has one electron in the 3d sub-shell and copper (29) has the configuration . . . $3s^2$ $3p^6$ $3d^{10}$ $4s^1$, i.e., the 3d sub-shell is full. Atoms with atomic numbers 22 to 28 have various occupation levels of the sub-shell. The elements from scandium to copper (which, as Cu^{++}, has an incomplete sub-shell) are known as the *first transition series* and the neutral atoms, or free ions, or both will be associated with a permanent magnetic dipole moment. This series is occasionally referred to as the iron group (iron has atomic number 26) but this name can be misleading.

There is also a second transition series running from yttrium (39) to silver (47) which has an incomplete 4d sub-shell. The *lanthanide series* consists of atoms which have an incomplete 4f sub-shell with 5s and 5p electron sub-shells complete outside it. A third transition series is deficient in 5d electrons and the actinide series lacks a complete 5f sub-shell. An element belonging to any of these five series may possess a non-zero magnetic moment when it occurs in a solid.

A typical paramagnetic salt is chrome alum, $CrK(SO_4)_2 . 12H_2O$. The magnetic moment is associated with chromium which occurs as the Cr^{3+} ion (. . . $3s^2$ $3p^6$ $3d^3$).† The other atoms give only a diamagnetic contribution to the susceptibility. Chrome alum obeys Curie's law quite accurately down to about 2 K. Below this temperature the susceptibility is complicated not only by the interactions between the chromium ions but also by the effect of the crystal field (see p. 152).

The equations derived in the last two sections were, strictly speaking, for a gas or a magnetically very dilute solid, for all interaction between the dipoles was ignored. The field acting on a dipole was assumed to be just the applied field. Generally, in the paramagnetic region, solids obey a modified Curie-type law known as a Curie–Weiss law

†It is assumed that the electronic structure of the chromium ion in the solid is very similar to that of the free Cr^{3+} ion. An atomic standpoint is used here but it is perhaps worth mentioning that very accurate calculations of the properties of the d-states would have to be done using some kind of tight-binding model as explained in the previous chapter.

$$\chi = \frac{C}{T - \theta} \quad \text{where} \quad T > \theta.$$

θ can be positive or negative and is associated with an internal ordering of the dipoles at low temperatures (see the sections on ferromagnetism and antiferromagnetism).

Calculation of the Curie constant

The expression for the Curie constant is

$$C = \frac{Ng^2 \beta^2 J(J+1)\mu_0}{3k_B}$$

and before we can evaluate this we will need to know J and g. To find J for a free atom or ion we use Hund's rules.

Hund's rules. These tell us how to calculate S, L, and J for an incompletely filled sub-shell. The first rule states that

1. The electron spins add to give the maximum possible S consistent with the Pauli principle.

The physical explanation for this rule depends on the behaviour of the 'exchange' terms in the expression for the total energy of the atomic system. Suppose we have two electrons and two electron states ψ_a and ψ_b. The total wavefunction, for particles which obey the exclusion principle and have half-integral spin, must be antisymmetric. Therefore, we can write

$$\Psi = \psi_a(r_1, \theta_1, \phi_1, s_1)\psi_b(r_2, \theta_2, \phi, s_2) - \psi_b(r_1, \theta_1, \phi_1, s_1)\psi_a(r_2, \theta_2, \phi_2, s_2).$$

s is a variable with two values, 1 and -1, and has the following properties:

$$\alpha(1) = 1, \quad \alpha(-1) = 0, \quad \beta(1) = 0, \quad \beta(-1) = 1$$

so

$$\sum_{s=\pm 1} \alpha(s)^2 = 1, \quad \sum_{s=\pm 1} \beta(s)^2 = 1, \quad \sum_{s=\pm 1} \alpha(s)\beta(s) = \sum_{s=\pm 1} \beta(s)\alpha(s) = 0.$$

There is a term in the Hamiltonian of the system $= 1/r_{12}$ which represents the Coulomb interaction between the two negatively charged electrons. The expectation value for the energy $\langle \Psi | H | \Psi \rangle$ will then contain terms

$$-\sum_{s_1}\sum_{s_2}\iint \psi_a(r_1,\theta_1,\phi_1,s_1)\psi_b(r_2,\theta_2,\phi_2,s_2) \frac{1}{r_{12}} \psi_b(r_1,\theta_1,\phi_1,s_1)\psi_a(r_2,\theta_2,\phi_2,s_2)\, d\tau_1\, d\tau_2$$

$$-\sum_{s_1}\sum_{s_2}\iint \psi_b(r_1,\theta_1,\phi_1,s_1)\psi_a(r_2,\theta_2,\phi_2,s_2) \frac{1}{r_{12}} \psi_a(r_1,\theta_1,\phi_1,s_1)\psi_b(r_2,\theta_2,\phi_2,s_2)\, d\tau_1\, d\tau_2$$

$$= -2\sum_{s_1}\sum_{s_2}\iint \psi_a(r_1,\theta_1,\phi_1,s_1)\psi_b(r_2,\theta_2,\phi_2,s_2) \frac{1}{r_{12}} \psi_b(r_1,\theta_1,\phi_1,s_1)\psi_a(r_2,\theta_2,\phi_2,s_2)\, d\tau_1\, d\tau_2$$

$$= \begin{cases} 0 \\ \text{or} \\ -\mathscr{J}_{ab} \text{ (where } \mathscr{J}_{ab} \text{ is the exchange energy† between } \psi_a \\ \quad \text{and } \psi_b\text{).} \end{cases}$$

This term is non-zero only when the spin parts of ψ_a and ψ_b are the same. If $\psi_a = \varphi_a\alpha(s)$ which corresponds to 'spin up' and $\psi_b = \varphi_b\beta(s)$ which corresponds to 'spin down', then Σ_s will be zero and the exchange term will be zero regardless of the integration over spatial coordinates. When both spins are up or both are down $\Sigma_s = 1$ and the exchange energy depends on the spatial integration. If ψ_a and ψ_b are *within the same atom* then $\mathscr{J}_{ab}$ is always postive although it will vary depending on the orbital states represented by φ_a and φ_b. Where $\mathscr{J}_{ab}$ is positive, the energy of the state with parallel spins is lower than the state with antiparallel spins (and the same φ_a and φ_b).

Let us take as an example Cr^{3+} (. . . 3s $3p^6$ $3d^3$). There are three electrons in the d-sub-shell which is capable of holding ten electrons.

†There is no classical explanation of exchange energy. It is not due to the fact that electron 1 swops position with electron 2. Indeed it occurs simply because the expression for Ψ takes into account the fact that the electrons are indistinguishable – to swop electron 1 with electron 2 is to leave the situation unchanged.

The orbital states are degenerate, i.e., the five values of m_l (2,1,0,–1,–2,) correspond to five φs with the same orbital energy. The three electrons will then each go into one of these states with spin (let us say) up. Two electrons cannot go into the same state with parallel spins for this would contradict the Pauli principle. So far we know nothing about their arrangement within these states except that three of the five orbital states will be singly occupied. $S = \frac{1}{2} + \frac{1}{2} + \frac{1}{2} = \frac{3}{2}$. (If there had been six electrons in the d-sub-shell then two would have had to pair up leaving $S = \frac{1}{2} + \frac{1}{2} + \frac{1}{2} + \frac{1}{2} = 2$.) The effect of the spin–spin interaction is to give a vector $\boldsymbol{S}$ (for Cr^{3+} of magnitude $\{\frac{3}{2}(\frac{3}{2} + 1)\}^{1/2}$) about which the s's precess.

2. The electron orbital angular momenta add to give the maximum possible L which is consistent with (1).

There are terms in the expression $\langle \Psi | H | \Psi \rangle$ which are equivalent to the electrostatic repulsion of the electron charge clouds. Such electrostatic effects cause an interaction between the orbital states and make the energy of an atomic state a minimum when rule (2) is obeyed.

In our example of Cr^{3+}, m_l can be 0, ±1, ±2 and the maximum value of M_L that we can produce is by filling the states $m_l = 2$, $m_l = 1$, and $m_l = 0$. Then $M_L = 2 + 1 + 0 = 3$ and corresponds to $L = 3$. The effect of the orbit–orbit interaction is to give a vector $\boldsymbol{L}$ of magnitude $\{3(3 + 1)\}^{1/2}$ about which the ls precess.

There is then a further (spin–orbit) interaction between $\boldsymbol{S}$ and $\boldsymbol{L}$ which is an interaction of the magnetic dipoles associated with the spin and with the orbital angular momentum. As a consequence we have a quantum number J and the last rule states that:

3. For a sub-shell less than half full, $J = L - S$.
   For a sub-shell more than half full, $J = L + S$.

The Cr^{3+} ion has a shell less than half filled so $J = L - S = 3 - \frac{3}{2} = \frac{3}{2}$. The spin–orbit interaction gives us a vector $\boldsymbol{J}$ (magnitude $\{\frac{3}{2}(\frac{3}{2} + 1)\}^{1/2}$) about which $\boldsymbol{L}$ and $\boldsymbol{S}$ precess. A summary of the calculation of J for Cr^{3+} is shown in Fig. 6.2.

Now we have managed to find J we still need to know how to derive the Landé g-factor before we can calculate the Curie constant.

The calculation of the g-factor. Figure 6.3(a) shows how the vector $\boldsymbol{J}$ is made up from $\boldsymbol{L}$ and $\boldsymbol{S}$; (b) shows how the total magnetic moment $\boldsymbol{m}_{tot}$ is the sum of $\boldsymbol{m}_{spin}$ and $\boldsymbol{m}_{orbit}$. Now, $m_{spin} = -g_s\beta\{S(S + 1)\}^{1/2}$ and g_s, the g-factor due to spin, is 2. Also $m_{orbit} = -g_l\beta\{L(L + 1)\}^{1/2}$ where g_l,

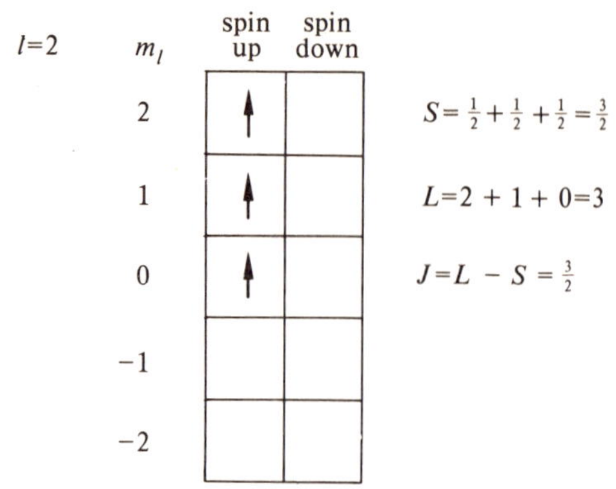

Fig. 6.2. The calculation of J for Cr^{3+}.

the g-factor due to orbital motion only, is 1. β is the Bohr magneton. By definition, the projection of $\boldsymbol{m}_{tot}$ upon the J-direction = m = $-g\beta\{J(J+1)\}^{1/2}$ and this is equal to

$$-2\beta\{S(S+1)\}^{1/2}\cos\gamma - \beta\{L(L+1)\}^{1/2}\cos\alpha.$$

It remains to calculate $\cos\gamma$ and $\cos\alpha$. From the first diagram:

$$\sin\alpha = x/\{L(L+1)\}^{1/2}, \quad 1-\cos^2\alpha = x^2/\{L(L+1)\},$$
$$\sin\gamma = x/\{S(S+1)\}^{1/2}, \quad 1-\cos^2\gamma = x^2/\{S(S+1)\}.$$

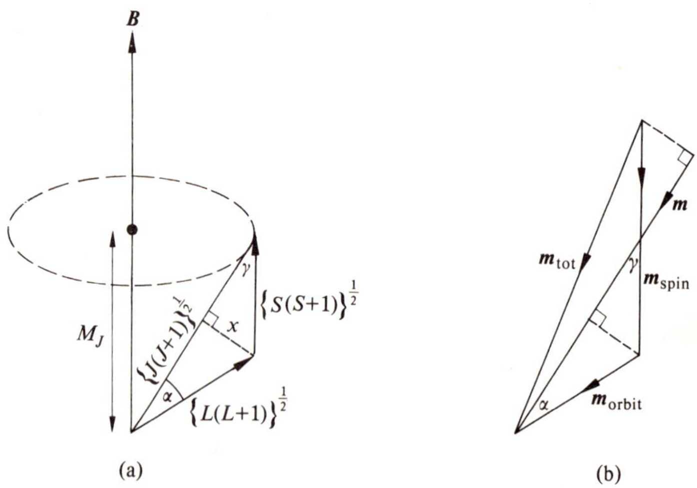

Fig. 6.3. (a) L and S add to give J. (b) $\boldsymbol{m}_{spin}$ + $\boldsymbol{m}_{orbit}$ = $\boldsymbol{m}_{tot}$. $\boldsymbol{m}_{tot}$ will precess about $\boldsymbol{m}$. The component perpendicular to $\boldsymbol{m}$ will average to zero.

Therefore,

$$L(L+1)(1-\cos^2\alpha) = S(S+1)(1-\cos^2\gamma).$$

Also,

$$\{S(S+1)\}^{1/2}\cos\gamma + \{L(L+1)\}^{1/2}\cos\alpha = \{J(J+1)\}^{1/2}.$$

The solution of these two equations yields

$$\cos\gamma = \frac{J(J+1) - L(L+1) + S(S+1)}{2\{S(S+1)J(J+1)\}^{1/2}}$$

and

$$\cos\alpha = \frac{J(J+1) + L(L+1) - S(S+1)}{2\{J(J+1)L(L+1)\}^{1/2}}.$$

$$\begin{aligned} m &= -g\beta\{J(J+1)\}^{1/2} \\ &= -2\beta\{S(S+1)\}^{1/2}\frac{J(J+1) - L(L+1) + S(S+1)}{2\{S(S+1)J(J+1)\}^{1/2}} \\ &\quad -\beta\{L(L+1)\}^{1/2}\frac{J(J+1) + L(L+1) - S(S+1)}{2\{L(L+1)J(J+1)\}^{1/2}} \end{aligned}$$

gives

$$g = \frac{3}{2} + \frac{S(S+1) - L(L+1)}{2J(J+1)}.$$

There are two situations, when $J = S$ and when $J = L$, in which we can easily check the correctness of our expression for g. When $J = S$, $L = 0$, and $g = 2$. When $J = L$, $S = 0$ and $g = 1$. These values are just what we need, g_s and g_l respectively.

The lanthanides

You have seen how to calculate J, g, and hence C for a free ion with an incomplete sub-shell. Do we get good agreement with experiment when we use these formulae? In the case of magnetically dilute salts of the lanthanides at approximately room temperature and above, we do. This group of elements have an incomplete 4f sub-shell outside which are the complete 5s and 5p sub-shells. Because the 4f sub-shell is fairly deep within the atom, the $\boldsymbol{L}$–$\boldsymbol{S}$ coupling is not split by the crystal field and J is a good quantum number. However, there is an interaction between the crystal field and the orbital angular momentum, which acts as a perturbation upon the J-states and which will remove some of the degeneracy of the $2J + 1$ energy levels. They are otherwise degenerate in the absence of an applied magnetic field. The total splitting corresponds to a few hundred absolute degrees of temperature, but when all the energy levels are fairly well populated (at about room temperature and above) the susceptibility is not very sensitive to the splitting and Curie's law is obeyed fairly closely (in magnetically dilute solids) with $C = \mu_0 g^2 \beta^2 J(J + 1)/3k_B$. At lower temperatures the susceptibility will not be equal to C/T and may be very anisotropic, i.e., have different values along different crystal directions.

The first transition series

This series of elements, including iron and chromium, are paramagnetic because their 3d sub-shell is incomplete. This 3d sub-shell is not buried in the interior of the ion as is the 4f sub-shell in the ions of the lanthanides. The ions behave very differently in the solid than when they are free. When experimental measurements on the paramagnetism of magnetically dilute salts containing elements of the first transition series are compared with the theoretical calculations, it is found that comparison of the measured Curie constant with $g^2 \beta^2 J(J + 1)\mu_0/3k_B$ shows poor agreement. A much better correspondence is obtained by using $g_s{}^2 \beta^2 S(S + 1)\mu_0/3k_B$. The explanation for this behaviour lies in the effect of the crystal field upon the energy states of the ion.

When the crystal field is strong in comparison with the spin–orbit interaction, the coupling is broken. J is no longer a good quantum number. The magnetic moment now depends on the separate contributions from the spin and from the angular momentum. When the crystal field, although strong enough to break the $\boldsymbol{L}$–$\boldsymbol{S}$ coupling, is weak in comparison with the spin–spin interaction, Hund's first rule is still obeyed and the contribution to the magnetic moment from the spin is $-\beta g_s S(S + 1)$. However, the crystal field has a further effect upon the orbital wavefunctions.

We can consider what happens in the case of an octahedral field (see Fig. 6.4(a)). The original five-fold degenerate orbits are split into a low-energy triplet of states with t_{2g} symmetry, sometimes known as dϵ states, and a high-energy doublet of states with e_g symmetry, sometimes known as dγ states. (Fig. 6.4(b).) A magnetic field can split the triplet levels. When there are 3, 4, 5, 8, or 9 d-electrons available, then the ground state energy of the ions will not be changed by the presence of a magnetic field and there will be no orbital contribution to the magnetic moment or susceptibility. We must take into account, as a perturbation, the spin–orbit effects. It is not true to say that the spin–orbit coupling is completely broken. The perturbation is larger for the heavier elements towards the end of the group. The effective value of g must vary slightly from 2 in order to make theory and experiment agree. Hund's third rule tells us that when the shell is less than half filled $J = L - S$ or the orbital angular momentum opposes the spin. In this case we should expect g to be less than 2. When $J = L + S$, for a more than half-filled sub-shell, we should expect the effective magnetic moment to be increased and therefore that we should need a g-factor greater than 2. (There is a further effect due to the remaining spin–orbit coupling. The crystal field, acting through the coupling, may split the degenerate S levels. The splitting is small – corresponding to $k_B T$ where T is of the

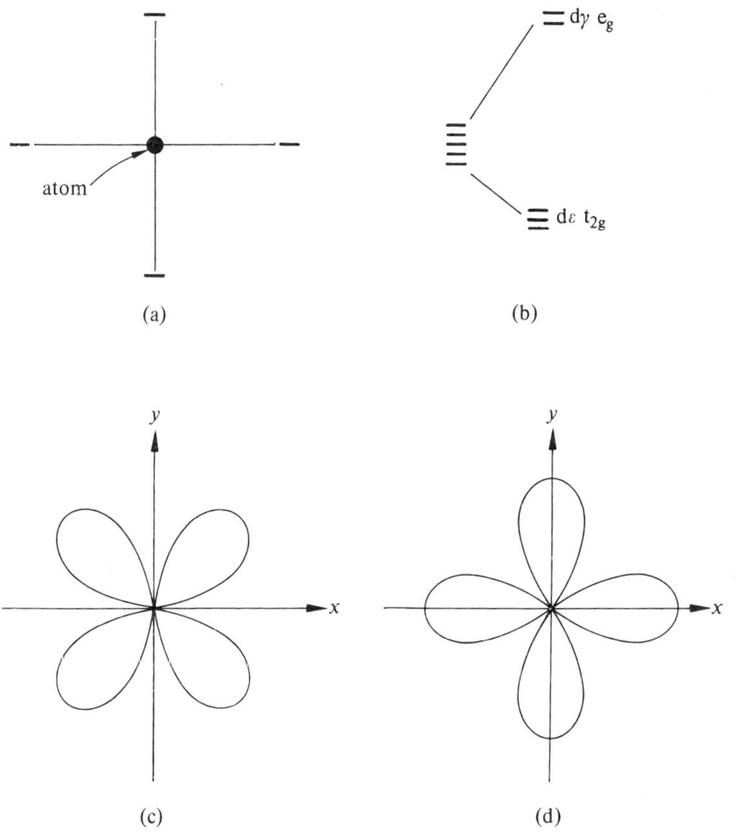

Fig. 6.4. (a) An octahedral field. (b) The splitting of the orbital energies. (c) Electron density when $\varphi_3 = xyf(r)$. (d) Electron density when $\varphi_2 = (x^2 - y^2)f(r)$.

order of 1 K – so that it can only affect the susceptibility at low temperatures.)

When there are 1, 2, 6, or 7 electrons in d-orbitals, then the ground state energy splits into three levels when a magnetic field is applied. L is 1 and there can be a first-order contribution to the susceptibility.

The octahedral crystal field shown in Fig. 6.4(a) acts as a perturbation upon the atomic energy levels. It will be of the general form

$$eV = Au(x^2) + Au(y^2) + Au(z^2)$$

where $u(x^2)$ is an even function of x, etc. The usual (complex) magnetic d-states for which the expectation values of L_z ($= -i\hbar\partial/\partial\varphi_x$ = the angular momentum operator for the z-direction) are $2\hbar$, $\hbar$, 0, $-\hbar$, and $-2\hbar$ are not of much help to us here for they are not diagonal to the perturbation. $\langle\varphi_i' \mid eV \mid \varphi_j'\rangle$ is not necessarily zero when $i \neq j$. We need to use real, non-magnetic d-states which are linear combinations of the magnetic d-states and for which $\langle\varphi \mid L_z \mid \varphi\rangle = 0$. They are of the form

$$\varphi_1 = \frac{1}{\sqrt{3}}(3z^2 - r^2)f(r) \qquad \text{depends on } \varphi'(m_l = 0)$$

$$\varphi_2 = (x^2 - y^2)f(r) \qquad \text{depends on } \varphi'(m_l = 2) + \varphi'(m_l = -2)$$

$$\varphi_3 = xyf(r) \qquad \text{depends on } \varphi'(m_l = 2) - \varphi'(m_l = -2)$$

$$\varphi_4 = yzf(r) \qquad \text{depends on } \varphi'(m_l = 1) + \varphi'(m_l = -1)$$

$$\varphi_5 = zxf(r) \qquad \text{depends on } \varphi'(m_l = 1) - \varphi'(m_l = -1).$$

Now $\langle\varphi_i \mid eV \mid \varphi_j\rangle = 0$ if $i \neq j$.

The new energy levels are given by the diagonal terms $\langle\varphi_i \mid eV \mid \varphi_i\rangle$. Because eV is symmetrical in x, y, and z

$$\langle xyf(r) \mid eV \mid xyf(r)\rangle = \langle yzf(r) \mid eV \mid yzf(r)\rangle = \langle zxf(r) \mid eV \mid zxf(r)\rangle.$$

In other words, the energy levels of φ_3, φ_4, and φ_5 are all changed equally. The change in energy of state φ_1 is

$$\left\langle(3z^2 - r^2)\frac{f(r)}{\sqrt{3}} \mid eV \mid (3z^2 - r^2)f(r)\right\rangle, \quad \text{and with } 3z^2 - r^2 = 2z^2 - x^2 - y^2$$

$$= \iiint dx\, dy\, dz \frac{f(r)}{3} eV\,(4z^4 + x^4 + y^4 + 2x^2y^2 - 4z^2x^2 - 4z^2y^2).$$

Also, the change in energy of state φ_2 is

$$\langle(x^2 - y^2)f(r) \mid eV \mid (x^2 - y^2)f(r)\rangle$$
$$= \iiint dx\, dy\, dz\, f^2(r)\, eV\,(x^4 + y^4 - 2x^2y^2).$$

But, since

$$\iiint dx\, dy\, dz\, f^2(r) eV x^4 = \iiint dx\, dy\, dz\, f^2(r) eV y^4 = \iiint dx\, dy\, dz\, f^2(r) eV z^4$$

and

$$\iiint dx\, dy\, dz\, f^2(r) eV x^2y^2 = \iiint dx\, dy\, dz\, f^2(r) eV y^2z^2$$
$$= \iiint dx\, dy\, dz\, f^2(r) eV z^2x^2$$

(because eV is symmetric in x, y, and z), the energies of φ_1 and of φ_2 are perturbed to the same extent.

We now have a triplet (φ_3, φ_4, φ_5) and a doublet (φ_1, φ_2). Which are the low-energy states? Figure 6.4(c) shows φ_3 and (d) shows φ_2. It is clear that φ_3 with the electron distribution between the axes and well away from the negative charges will have the lower energy. The triplet has the lower energy.

A magnetic field applied to these new energy levels as a perturbation will cause mixing of φ_1 with φ_2 and of φ_3, φ_4, and φ_5. The mixing of φ_3, φ_4, and φ_5 can yield φ_3, $\varphi'(m_l = 1)$, and $\varphi'(m_l = -1)$ and so the field will split the three states and they will no longer be degenerate.

Strong crystal field and breakdown of spin–spin coupling

There are a few substances containing atoms of the first transition series, such as the complex cyanides, and many containing elements of the second and third transition series, in which the bonding is not purely ionic but is partly covalent. In such cases the effective crystal field is very strong – strong enough to overcome the spin–spin coupling which is the basis for Hund's first rule. The d-electrons simply fill up the low-energy dϵ states first, and then the dγ states (assuming octahedral field). The situations with four, seven, and eight electrons are shown in Fig. 6.5.

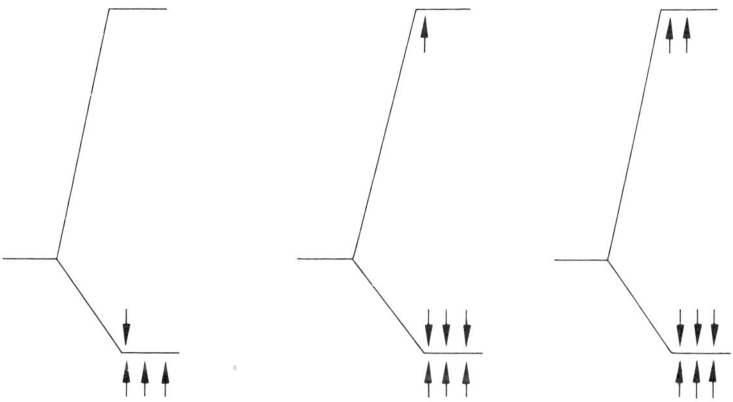

Fig. 6.5. Electron states in a strong crystal field (a) d^4, (b) d^7, (c) d^8.

Two experimental techniques making use of paramagnetic behaviour

Adiabatic demagnetization. Low temperatures can be produced by the technique of adiabatic demagnetization. A paramagnetic solid is used. The energy levels of the dipole system will be slightly split in the absence of an applied magnetic field and its entropy will be a function of the temperature T_1. When a strong magnetic field B_0 is applied to the solid, the dipoles will tend to align themselves with the field. The system will become more ordered than it was before the field was applied. The entropy with $B = B_0$, $T = T_1$ will be lower than $S(B = 0, T = T_1)$ and will in fact correspond to $S(B = 0, T = T_2)$ where T_2 is a lower temperature than T_1. Corresponding with the decrease in entropy, a certain amount of energy will be passed from the dipole system to the lattice. If the solid is kept at constant temperature T_1 while the field is applied, heat will have to be removed from the solid. Then the solid is thermally isolated from its surroundings and the magnetic field B_0 is reduced to zero. Ignoring the entropy of the lattice vibrations which will be small in the temperature regions under consideration (1 K downwards), the entropy remains constant so

$$S(B = B_0, T = T_1) \rightarrow S(B = 0, T = T_{\text{final}}) = S(B = 0, T = T_2)$$

and the final temperature is T_2 which has already been shown to be less than T_1.

Electron paramagnetic resonance (e.p.r.). The energy levels of a magnetic electron-state are separated by $g\beta B$ in a magnetic field B. If electromagnetic radiation of frequency ν such that

$$h\nu = g\beta B$$

is available, then there may be absorption of a photon and transition to a higher energy state. Usually, in this type of resonance experiment, ν is kept fixed at a frequency corresponding to a wavelength of about 3 cm and B (approximately 0.1 to 1 tesla or weber m^{-2}) is swept until resonance absorption occurs. (For experimental detail see, for example, McLauchlan[1].)

This technique may be used to study colour centres in solids. For example, the F-centre in alkali halide crystals consists of an unpaired electron trapped at a vacant negative-ion site. Electron resonance experiments show that g is very nearly 2, i.e., that the F-centre wavefunction is reasonably well described by a simple s-state function.

Studies by resonance experiments of paramagnetic ions in crystals can yield much valuable information about the strengths and directions of crystal fields. The experimental data is easier to interpret if the resonance lines are narrow. This will be the case if the lifetime of the excited state is long.† When the paramagnetic ion is in an S-state (true for all ions of the first transition series and for the odd ion of the 4f group), the lattice vibrations, acting via the crystal field, have very little

†The Uncertainty Principle states that $\Delta E \times \Delta t$ has a minimum value $\sim\hbar$.

effect on the magnetic moment and the lifetime is long. Some ions of the 4f group (lanthanides) have dipole moments with sufficiently strong interaction, through $\boldsymbol{L}$, with the crystal field to make the resonance lines too broad to deal with except at very low temperatures. The 4f sub-shell is deep enough within the ion for the crystal field to be unable to break the $\boldsymbol{L}$–$\boldsymbol{S}$ coupling. But it will act as a small perturbation and have much stronger contact with J than it has with $\boldsymbol{S}$ for the 3d group. There the crystal field has very strong control over the orbital states but the magnetic moment associated with $\boldsymbol{S}$ is almost free from contact with an $\boldsymbol{L}$. The technique of electron paramagnetic resonance is therefore easiest to use when the magnetic moment is associated with electron spin only and g is 2.

Ferromagnetism – the Weiss theory

A few materials, notably iron, can be spontaneously magnetized at ordinary temperatures. They behave like 'magnets' and are magnetized in the absence of an external field. In order to explain such properties P. Weiss introduced (in 1907) the idea of an internal field. Suppose that the field 'seen' by an atomic dipole is not simply the applied field B_0, but is $B_0 + B_{int}$ where B_{int} is the internal field and is assumed to be dependent on the magnetization.

$$B_{int} = \lambda M.$$

When† $B = B_0 + \lambda M$ is such that $g\beta B/k_B T$ is small then we might expect to be able to use eq. (6.4) with $y = Jg\beta B/k_B T = Jg\beta(B_0 + \lambda M)/k_B T$. Equation (6.4) becomes

$$\chi \equiv \frac{M}{B_0}\mu_0 = Ng\beta J \cdot \frac{(J+1)}{J} \cdot \frac{\mu_0}{B_0} \cdot \frac{Jg\beta(B_0 + \lambda M)}{3k_B T}$$

$$= \frac{Ng^2\beta^2 J(J+1)}{3k_B T}\mu_0 \cdot \left(1 + \frac{\lambda M}{B_0}\right).$$

$$\chi = \frac{C}{T}\left(1 + \frac{\lambda\chi}{\mu_0}\right) \quad \text{or} \quad \chi = \frac{C}{T - \theta} \tag{6.5}$$

where $\theta = C\lambda/\mu_0$. θ is called the Weiss constant or the Curie temperature. Provided y is small the material will have paramagnetic behaviour obeying the Curie–Weiss law. Assuming that eq. (6.5) applies at $T = \theta$,

†In effect this formula implies that M is in the same direction as B_0. This certainly applies in the paramagnetic region. At low temperatures, for a single domain (see later), this may not necessarily be true. The spontaneous magnetization will tend to align with certain crystal directions.

what do we predict about the susceptibility? When $T = \theta$, χ is infinite. This means that $\mu_0 M/B_0$ is infinite. M has a maximum possible value of $NJg\beta$, when all the dipoles are aligned parallel, and therefore an infinite value of χ can only mean that M remains non-zero when B_0 is zero. The solid will exhibit spontaneous magnetization.

If we want to do calculations below the high-temperature region in which eq. (6.5) is valid, it is necessary to make use of eq. (6.3). $\bar{m} = g\beta J Br(y)$ where $Br(y)$ is the Brillouin function and $y = Jg\beta B/k_B T$. We have

$$\frac{N\bar{m}}{NJg\beta} = \frac{M_{sat}^T}{M_{sat}^0} = Br\left\{\frac{M_{sat}^0}{Nk_B T}(B_0 + \lambda M)\right\}$$

where M_{sat}^0 and M_{sat}^T are, respectively, the maximum possible values (see next section) of the magnetization at 0 K and T K.

When $B_0 = 0$, no applied field,

$$\frac{M_{sat}^T}{M_{sat}^0} = Br\left(\frac{M_{sat}^0}{Nk_B T}\lambda M_{sat}^T\right).$$

The equation can be solved, for each value of temperature, to find the value of M_{sat}^T – the magnetization in zero field. The results found are shown in Fig. 6.6. At absolute zero $M_{sat}^T = M_{sat}^0$, and at higher temperatures M_{sat}^T falls, to reach zero at the Curie temperature. (For $T < \theta$ there are always two solutions of the equation – one corresponding to $M_{sat}^T/M_{sat}^0 = 0$ and the other to $0 < M_{sat}^T/M_{sat}^0 < 1$. It can be shown that the solution $M_{sat}^T/M_{sat}^0 = 0$ represents an unstable

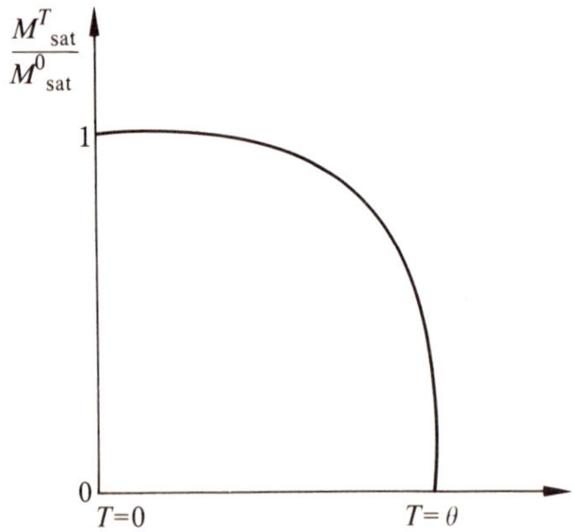

Fig. 6.6. Values of spontaneous magnetization of ferromagnetic material at temperatures below the Curie point.

situation.) Note that when $T = \theta, M_{sat}^T = 0$ in zero field: so we expect, on this simple theory, that the Curie–Weiss law should apply for all $T > \theta$ provided only that B_0 is not too large.

Domains and hysteresis

One of the problems with the theory so far is that it predicts that all ferromagnetic materials should be spontaneously magnetized at temperatures below the Curie point. The Curie points of iron, nickel, and cobalt are shown in Table 6.1. Not all pieces of iron are magnetized at room temperature. They are usually in a 'demagnetized' state. They will become magnetized when placed in a magnetic field and part of the magnetization may remain even when the field is removed. A field applied in the opposite direction can remove the magnetization (and so can mechanical shock). The magnetization curve (OPQR in Fig. 6.7) and the hysteresis loop (RSTUVWR in Fig. 6.7) of iron at room temperature are well known.

TABLE 6.1

	Curie temperature
Iron	1043 K
Cobalt	~1390 K
Nickel	~630 K

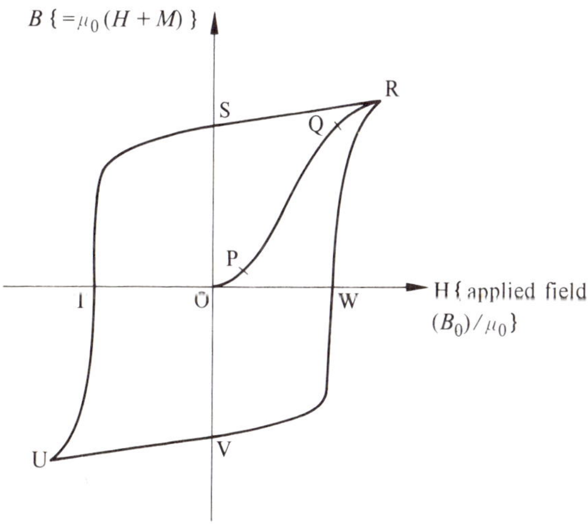

Fig. 6.7. Hysteresis loop for a ferromagnetic material.

In order to explain the macroscopic behaviour of ferromagnetic materials Weiss introduced the idea of *domains*. These are tiny regions (since observed to be $\sim 10^{-6}$ to 10^{-2} cm^3) of the material in which the dipoles are aligned parallel and M is just that which would be expected from the theory, M^T_{sat}. But these domains may be arranged in such a way that the total magnetization seen from outside the specimen is zero. Look at Fig. 6.8. The first diagram represents a piece of iron which is not divided into domains. (b) represents a division into four domains. The triangular domains at the end of the specimen are known as closure domains. The specimen as a whole is now demagnetized and demagnetization is associated with a decrease in energy of the state. However, there are 'easy' and 'hard' directions of magnetization in crystals depending on which way the dipoles will have to point in relation to crystal axes. Our closure domains are magnetized in a hard direction. In (c) the volume of material magnetized in this

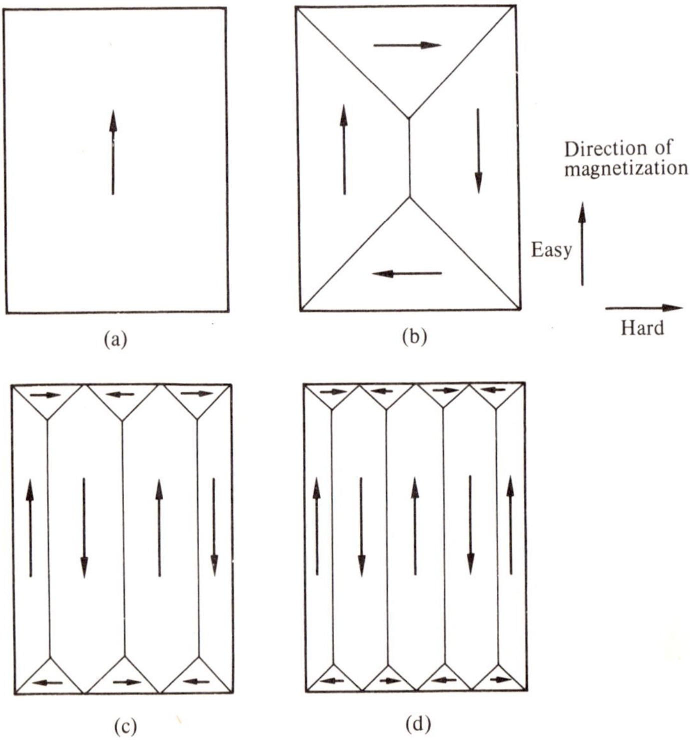

Fig. 6.8. Domains.

direction has decreased. This situation represents an even more energetically favourable state. Surely (d) will be better yet? Not necessarily. Energy is needed to maintain the domain boundaries or Bloch walls as they are known. This effect prevents the domains becoming too small. In fact, very small grains of iron must remain as single domains. (c) is then our stable state and is a consequence of a balance of the energies due to demagnetization, anisotropy (associated with hard and easy directions of magnetization), and the production of domain walls.

Now to explain the magnetization curve of Fig. 6.7. Originally the specimen is demagnetized, for the domains are arranged in such a way that there are an equal number of dipoles pointing in each direction and there is no net flux outside the specimen. When an external field is applied the first thing that happens is that domains with dipoles pointing in the field direction grow at the expense of other domains in which the dipoles are oriented in less favourable directions. The movements of the domain walls are reversible. If the field is switched off the magnetization will revert to zero. This behaviour takes place in the region OP of Fig. 6.7. When the field increases further the domain walls begin to move irreversibly (in small jumps which can be seen if the curves are measured accurately enough). If the field were now to be removed there would be some residual magnetization. Above the point Q on the magnetization curve, B and hence M is obviously increasing only with great difficulty. A large field is needed to change M by a small amount. What is happening here is that whole domains of dipoles facing against the field direction are rotating their direction of magnetization.

Weiss was able to use the concept of domains to account for some of the properties of macroscopic pieces of ferromagnetic substances. Domains can be observed under a microscope. First the surface of the ferromagnetic material is very carefully cleaned and polished. Then, very finely divided iron is scattered onto the surface. It collects along the domain boundaries on the surface of the crystal because it is there that the local fields are strongest. It is interesting that very small grains of ferromagnetic materials (less than about 10^{-6} cm in length) always occur as single domains, for the demagnetized state is no longer energetically favourable.

Experimental values and deviations from the simple theory

The experimental values of inverse susceptibility plotted against absolute temperature for a ferromagnetic metal give a curve similar to that shown in Fig. 6.9. At high temperature the metal obeys a Curie–Weiss relationship with a parameter θ_p known as the paramagnetic

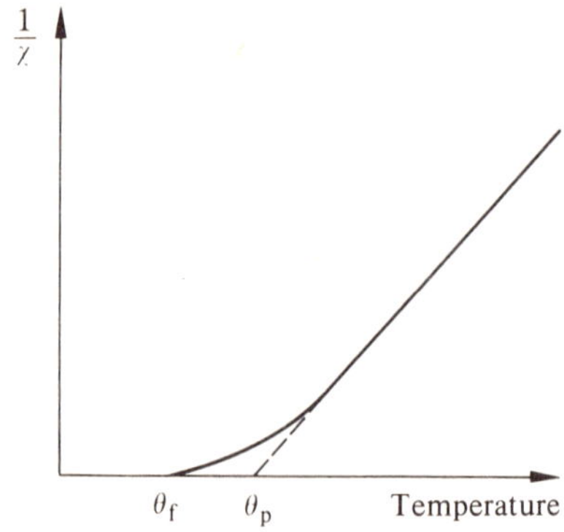

Fig. 6.9. Curve representing values of inverse susceptibility plotted against absolute temperature.

Curie temperature. At temperatures nearer to the Curie point the simple linear relationship between χ and T breaks down and the metal actually becomes ferromagnetic at a temperature θ_f known as the ferromagnetic Curie temperature (to distinguish it from θ_p) which is lower than θ_p. The simple theory of Weiss does not distinguish between a paramagnetic and a ferromagnetic Curie temperature.

In order to calculate $C = Ng^2\beta^2 J(J+1)\mu_0/3k_B$ and $M^0_{sat} = Ng\beta J$ to compare with experiment we need to know what to use for g and for J (or S). Dysprosium (Dy) is a ferromagnetic metal belonging to the lanthanide series. The atomic configuration is . . . $4f^{10}\ 5s^2\ 5p^6\ 6s^2$. If we assume that the occupancy of the 4f sub-shell remains the same in the metal and that J is a good quantum number, as it was when we were dealing with paramagnetic salts, then Hund's rules yield $S = 2$, $L = 6$, $J = 8$. The formula for g gives 5/4. Then, $gJ = 10$. Experiments measuring M^0_{sat} give the effective Bohr magneton number, $g\beta$, as 10.0. Here is very good agreement between experiment and the theory.

Unfortunately, the situation is more complicated when we have to try and deal with ferromagnetic substances containing the elements of the first transition series. When we were considering paramagnetic salts of these elements we could often deal with compounds that were sufficiently ionic for us to have a good idea of the *oxidation state* (the number of positive charges on the ion) of the magnetic element and hence of the number of electrons in the d sub-shell. Ferromagnetism seems to be *mainly* restricted to five metals (iron, cobalt, and nickel of the 3d group; gadolinium and dysprosium of the 4s group) and to alloys of these metals, with themselves or with arsenic, bismuth, and antimony. Mostly the bonding is metallic. There are only a few ferromagnetic insulators, such as europium sulphide.

The observed values of the effective magneton numbers of iron, cobalt, and nickel are shown in Table 6.2. Also shown in the table are the values calculated using S as a good quantum number and assuming that the occupancy of the d-states is the same as it is in the atom. The reason for the discrepancies can be given roughly as follows: The 4s band in the metal (the conduction band) overlaps the narrow 3d band and therefore some of the metals that had 4s states in the free atoms take up states in the 3d band when the atoms join together to form a metal. With S still a good quantum number and $g = 2$ we need on average 7.78 electrons in the d-states around each iron atom, 8.28 around each cobalt atom, and 9.39 around each nickel atom to reach agreement with experiment.

TABLE 6.2

	Atomic configuration	S	gS	Effective magneton number (observed).
Fe	$\ldots 3d^6\ 4s^2$	2	4	2.22
Co	$\ldots 3d^7\ 4s^2$	$\frac{3}{2}$	3	1.72
Ni	$\ldots 3d^8\ 4s^2$	1	2	0.61

Antiferromagnetism and ferrimagnetism

Although all magnetic solids will reach some kind of ordered state at low enough temperatures, simple ferromagnetism, with all the magnetic dipoles lined up parallel, is comparatively rare. Another possible ordered arrangement of dipoles is shown in Fig. 6.10. Here each dipole points in the opposite direction to that of all its nearest neighbours. We have effectively two sublattices – in each sublattice all the spins are arranged in parallel. A substance which behaves in this way is said to be antiferromagnetic. Evidence for this type of magnetic ordering is obtained from neutron diffraction experiments.

We can calculate the susceptibility in the high-temperature, paramagnetic region just as we did for a ferromagnetic substance, by using eq. (6.4). If there are $N/2$ atoms on *each* of the sublattices and if $M_A(T), M_B(T)$ are the magnetizations of the sublattices at temperature T

$$\chi = \chi_A + \chi_B = \frac{C}{2T}\frac{B}{B_0}$$

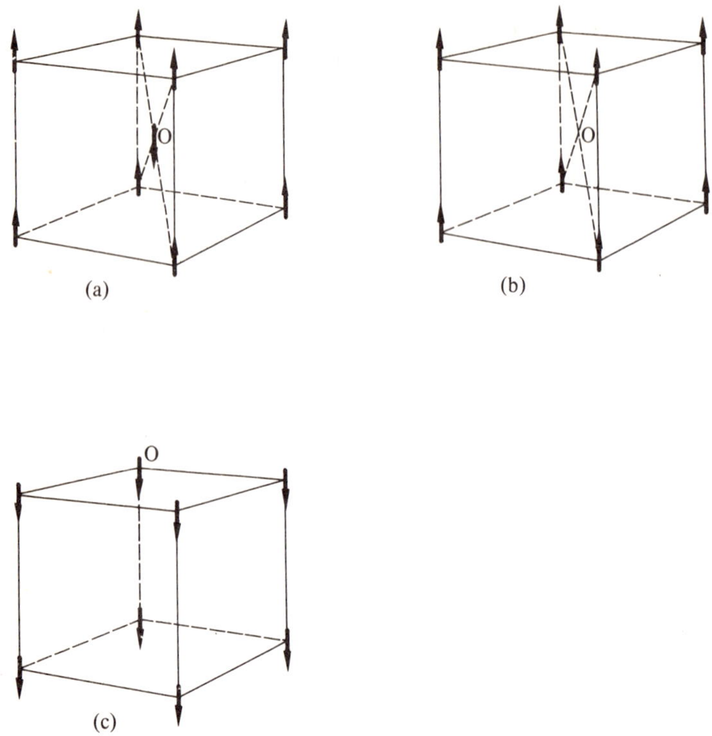

Fig. 6.10. (a) Magnetic element on b.c.c. lattice. (b) Simple-cubic sublattice with dipoles up. (c) Simple-cubic sublattice with dipoles down.

where

$$C = \frac{N g_s{}^2 \beta^2 S(S+1)\mu_0}{3k_B} \quad \text{(if } S \text{ is a good quantum number).}$$

Now, if we can say that each dipole has interactions only with its nearest neighbours, which are in the other sublattice, we can write the internal field acting on a dipole in sublattice A as $B_{int} = -\lambda M_B$.† Similarly, the internal field acting on a dipole in sublattice B is $B_{int} = -\lambda M_A$. The minus sign indicates that the exchange interaction (see next section) is such as to make the dipole point in the opposite

† Again we have the assumption, true above the Néel point, that M_A and M_B are colinear with B_0.

direction to its neighbours. Then

$$\chi = \chi_A + \chi_B = \frac{C}{2TB_0}(B_0 - \lambda M_B + B_0 - \lambda M_A)$$

$$= \frac{C}{2TB_0}\left(2B_0 - \lambda \frac{\chi}{\mu_0} B_0\right)$$

or

$$\chi = \frac{C}{T + \theta_p} \quad \text{where} \quad \theta_p = \frac{C\lambda}{2\mu_0}.$$

In the paramagnetic region the susceptibility obeys a Curie–Weiss law. Below a temperature θ_N known as the Néel temperature, M_A and M_B are non-zero in the absence of an external field. On this simple theory, involving only nearest neighbour interactions, the Weiss constant θ_p and the Néel temperature θ_N are identical. However, experiments show that θ_p and θ_N may vary considerably. This can be accounted for by improving the theory to include interactions with more distant neighbours which are on the same sublattice as the dipole in question. Note that $M_A = -M_B$ when T is less than θ_N and $B_0 = 0$ so there is no net spontaneous magnetization. The susceptibility is never infinite. It has a maximum at the Néel temperature and below this it is anisotropic (Fig. 6.11(a)).

The interactions between the magnetic atoms or ions will cause ordering into sublattices. The crystal field will tend to align the directions of magnetization along preferred crystal directions. We can easily calculate $\chi_\perp$, the susceptibility in a direction perpendicular to that of dipole alignment, for a two-sublattice model (see Fig. 6.11(b)).

$$\chi_\perp = \mu_0 \frac{M_A + M_B}{B_0}.$$

The field acting on each dipole of sublattice A is $B_0 - \lambda M_B$ and that on each dipole of sublattice B is $B_0 - \lambda M_A$. If B_0 (at right angles to $M_A(B_0 = 0)$ and to $M_B(B_0 = 0)$) is small, then we can write

$$M_A = M_A(B_0 = 0) + \delta M, \quad M_B = M_B(B_0 = 0) + \delta M$$

where δM is as shown on the diagram. Also, because M_A will align with the effective field acting on it

$$\tan\alpha = \frac{\delta M}{M_A(B_0 = 0)} = \frac{B_0 - \lambda\delta M}{\lambda M_A(B_0 = 0)}.$$

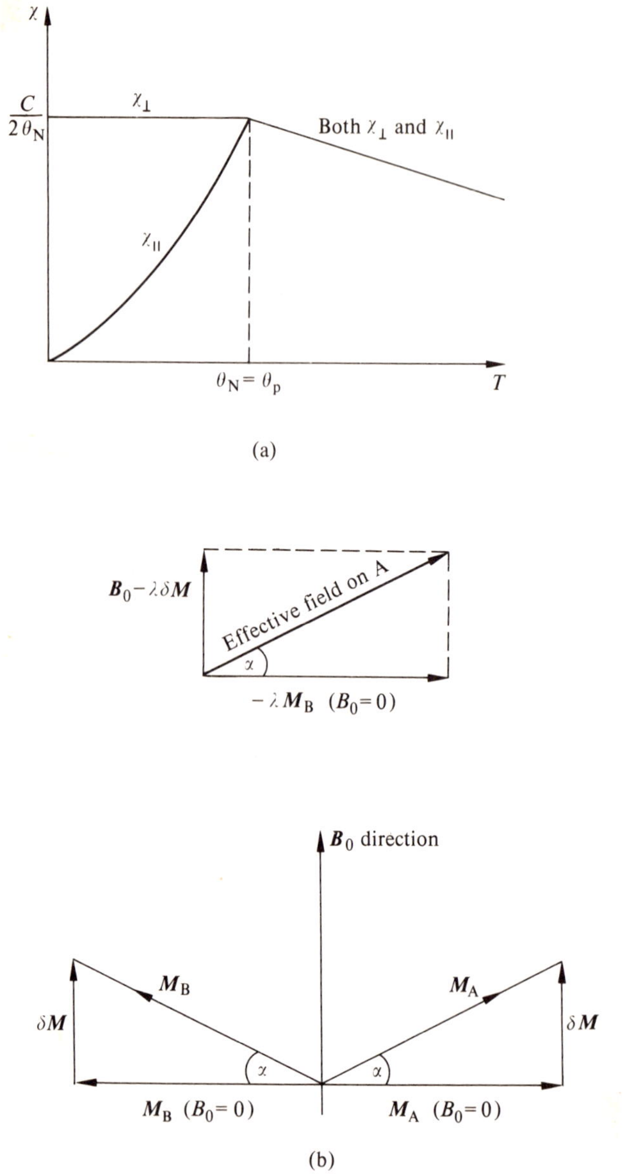

Fig. 6.11. (a) The susceptibility of an antiferromagnetic material calculated according to a simple theory. (b) Diagram corresponding to calculation of $\chi_\perp$ below the Néel temperature. α is shown fairly large for clarity.

Therefore

$$\delta M = \frac{B_0 - \lambda \delta M}{\lambda}$$

Then

$$\chi_\perp = \mu_0 \frac{M_A + M_B}{B_0} = \mu_0 \frac{M_A(B_0 = 0) + \delta M + M_B(B_0 = 0) + \delta M}{B_0}$$

$$= \mu_0 \frac{2\delta M}{B_0} = \frac{\mu_0}{\lambda} = \frac{C}{2\theta_N}.$$

This is independent of temperature and is equal to the value at the Néel temperature as calculated from $\chi = C/(T + \theta_N)$.

The calculation for susceptibility in a direction parallel to the alignment of the spins is much more complicated. However, one would expect it to drop to zero at the absolute zero of temperature for then all the spins would be aligned in their sublattices and the only way to change the value of the magnetization (zero in zero field) would be to 'turn over' one of the dipoles. A small field would not have enough strength to do this.

A few of the stronger antiferromagnetic materials are listed with their characteristic temperatures in Table 6.3.

TABLE 6.3

	θ_N	θ_P
MnO	116 K	610 K
FeO	198 K	570 K
CoO	291 K	330 K
MnTe	307 K	690 K

Ferrimagnetism is a variation on simple antiferromagnetism. The theory is based on a two sublattice model. The magnetic dipoles on sublattice A point in the opposite direction to those on sublattice B. The internal field is such as to cause a particular dipole to align itself antiparallel to its neighbours in the other sublattice. However, the total magnetic moment associated with sublattice A is different from that on B. This is because we have two different magnetic ions in the crystal. The result is that the net spontaneous magnetization below the Néel temperature is not zero but finite. (See Fig. 6.12.)

One class of ferrimagnets are the *ferrites* of general formula $M^{++}Fe_2^{+++}O_4^{---}$ where M is a member of the first transition series. An example of a ferrite is magnetite, Fe_3O_4.

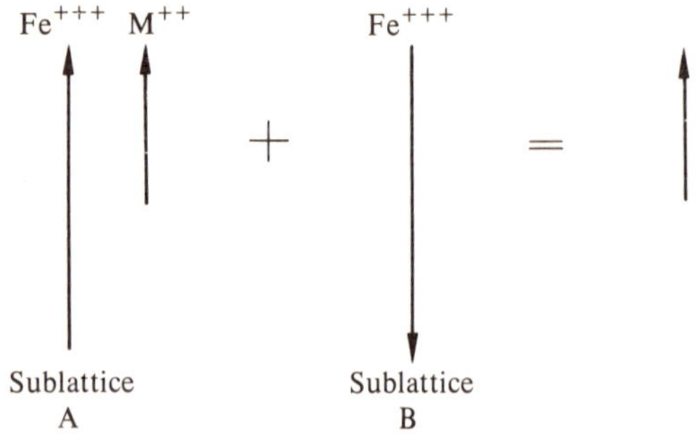

Fig. 6.12. Schematic representation of magnetization of sublattices in a ferrite at 0 K.

Another class of ferrimagnets are the *garnets* of general formula $M_3^{+++}Fe_5^{+++}O_{12}^{--}$ where M is usually a member of the lanthanide series.

Ferrimagnetic materials are of great importance in industry. They may have large magnetization at room temperature and yet they are electrical insulators. This makes them much more suitable than the ferromagnetic metals for use at high frequencies when eddy-current loss can be a serious problem.

There are yet other types of magnetic ordering at low temperatures. For example, the dipoles may be arranged in a spiral pattern. Or they may be similar in arrangement to a simple antiferromagnet except that the directions of magnetization of the sublattices may not be at exactly 180° to each other. These types of behaviour are only of rather rare academic interest.

The internal field

Weiss suggested a theory for ferromagnetism based on the concept of an internal field. He suggested that B_{int} be written as λM but offered no explanation of the forces involved. Simple (magnetic dipole)–(magnetic dipole) interactions are very weak and are unimportant at temperatures above about 1 K. To explain a transition temperature of 1000 K we need an energy of interaction of something like 1000 $k_B \approx$ 0.1 eV. B_{int} is then of the order of (0.1 eV)/β where β is the Bohr magneton. It turns out that B_{int} needs to be about 10^3 tesla. This is an enormous field – much bigger than those produced in the laboratory even now. It was not until 1928 that Heisenberg was able to offer a quantum mechanical explanation of this phenomenon.

It was explained in the section on Hund's rules that the energy of a state with electron spins parallel differs from that with electron spins antiparallel by $-\mathscr{I}_{ab}$. The spin–spin interaction can be written as a direct coupling between the spins of $E = -2\mathscr{I}_{ab}s_a \cdot s_b$. When s_a is parallel to s_b, $E = -2\mathscr{I}_{ab}\frac{1}{2} \times \frac{1}{2}$; when s_a is antiparallel to s_b, $E = +2\mathscr{I}_{ab}\frac{1}{2} \times \frac{1}{2}$. The difference in energies is just $\mathscr{I}_{ab}$.

The idea of a coupling between the spins can be extended to include coupling between an atom with total spin quantum number S_i and one with total spin S_j.

$$E = -2\mathscr{I}S_i \cdot S_j. \tag{6.6}$$

It must be realized that $\mathscr{I}$ is not necessarily positive. It will depend for both magnitude and sign on the distance apart of the atoms involved as well as on details of the atomic orbitals. Equation (6.6) is fine if S is a good quantum number as it is for the first transition series, but, when J is a quantum number, it must be modified. We need to know the projection of $\boldsymbol{S}$ upon $\boldsymbol{J}$. From Fig. 6.3 this can be seen to be

$$\{S(S+1)\}^{1/2} \cos\gamma = \{S(S+1)\}^{1/2} \frac{J(J+1) - L(L+1) + S(S+1)}{2\{S(S+1)J(J+1)\}^{1/2}}$$

$$= \{J(J+1)\}^{1/2}(g-1).$$

The energy of interaction can be written as

$$E = -2\mathscr{I}(g-1)^2 J_i \cdot J_j \tag{6.7}$$

I shall now work with eq. (6.7). When S is the good quantum number, substitute S for J and g_s for g in the following equations.

$$E = -2\mathscr{I}(g-1)^2 J_i \cdot J_j$$

$$= -2\mathscr{I}(g-1)^2 \frac{m_i}{g\beta} \cdot \frac{m_j}{g\beta}.$$

Now suppose that the dipole m_i interacts with n nearest neighbours of average $m_j = \bar{m}$. Then

$$E = -2\mathscr{I}(g-1)^2 \frac{m_i}{g\beta} \cdot \frac{n\bar{m}}{g\beta}.$$

If $\mathscr{I}$ is positive the internal field will be in the same direction as the magnetization and the exchange interactions will give rise to ferromagnetism. All the dipoles will tend to point in the same direction so $N\bar{m} = M$, and, as E can perfectly well be written as $-m_i \cdot B_{\text{int}}$,

$$-m_i \cdot B_{\text{int}} = -2\mathscr{I}(g-1)^2 \frac{m_i}{g\beta} \cdot \frac{M}{g\beta} \frac{n}{N}$$

$$B_{\text{int}} = \frac{2\mathscr{I}(g-1)^2}{g^2\beta^2} \frac{n}{N} M \quad \text{or} \quad \lambda = \frac{2\mathscr{I}(g-1)^2 n}{g^2\beta^2 N}.$$

If $\mathscr{I}$ is negative we will have an internal field which is in the opposite direction to the magnetization of the nearest neighbours. This will give rise to antiferromagnetism. If we use a two-sublattice model and restrict the interaction to the nearest neighbours, which will be on the other sublattice, then

$$B_{\text{int}} \text{ (acting on a dipole in A)} = \frac{2\mathscr{I}(g-1)^2 nM_{\text{B}}}{g^2\beta^2 (N/2)} \quad \text{or} \quad -\lambda M_{\text{B}}$$

where λ is positive. There are $N/2$ atoms (or ions) on each sublattice.

An exchange interaction, then, can in principle explain the existence of an effective internal field. But will it explain the magnitude of the effects involved? Within an atom the exchange interaction between two spins may be of the order of 1 eV and so it seems possible that a direct exchange interaction will explain the order of the Curie and Néel temperatures. But overlap of the d- or f-orbitals on neighbouring atoms is small and hence the direct exchange energies between them are also rather smaller than we would like. It seems unlikely that direct exchange interactions can explain strong ferromagnetism. There does not seem to be any totally accepted theory as to how the exchange mechanisms work but it appears possible that the interaction in metals may take place via the conduction electrons and that in insulators it may be due to the presence of some covalent bonding between the magnetic metal ions and the anions. The anion between two cations would then be involved in the exchange interaction.

Examples

1) 1968 Paper IVA

In a certain ferromagnetic material the transition temperature is 300 K and the magnetic carriers have spin $S = \frac{1}{2}$ and a magnetic moment of one Bohr magneton. Find the value of the internal field at low temperatures.

Answer

The internal field $B_{\text{int}} = \lambda M$. The constant λ is given by the relationship

$$\theta = \frac{\lambda C}{\mu_0}$$

where θ is the transition temperature and C the Curie constant, $\mu_0 Ng^2\beta^2 J(J+1)/3k_B$.

$$\theta = \lambda Ng^2\beta^2 J(J+1)/3k_B \quad \text{gives}$$

$$300 = \lambda N4\beta^2\ \tfrac{1}{2}\ \tfrac{3}{2}/3k_B$$

where I have written $g = 2$ and $J = S = \tfrac{1}{2}$.

At low temperatures M approaches $Ng\beta J$ – all the dipoles in line.

$$B = \lambda M = Ng\beta J = N2\beta\tfrac{1}{2}$$

$$= \frac{300 \times 3k_B}{3\beta} \quad \text{(substituting for } \lambda)$$

$$= \text{approx. } 450 \text{ T.}$$

(N.B. This is an *enormous* field.)

2) 1970 Paper IVA

The ferromagnet europium oxide has a Curie temperature of 70 K. The europium ion has $J = \tfrac{7}{2}$ and $g = 2$. Assuming the molecular field model, determine the ratio of the magnetization at 300 K in a field of 100 gauss (0.01 tesla), to that at 0 K.

Answer

The susceptibility of the oxide at temperatures well above the transition temperature can be written as

$$\mu_0 \frac{M}{B_0} = \frac{C}{T-\theta}$$

where $C = \mu_0 Ng^2\beta^2 J(J+1)/3k_B$ is the Curie constant and B_0 the applied field. With the transition temperature = 70 K,

$$M = \frac{Ng^2\beta^2 J(J+1)/3k_B}{T-70} \times B_0$$

At very low temperatures the saturation magnetization is just that which is obtained with all the dipoles aligned parallel.

$$M(0) = Ng\beta J.$$

Therefore

$$\frac{M(300)}{M(0)} = \frac{g\beta(J+1)/3k_B}{T-70} \times B_0 = \frac{2\beta\frac{9}{2} \times 0.01}{3k_B \times 230}$$

$$= \frac{0.03}{230} \times \frac{9.27 \times 10^{-24}}{1.38 \times 10^{-23}}$$

$$\sim 9 \times 10^{-5}$$

This small ratio is what one might expect – paramagnetism is much smaller than ferromagnetism.

3) 1971 Pure Physics Option I

A primary and a secondary coil are wound tightly around a long thin crystal of a dilute nickel salt with $g = 2$ which contains 10^{21} Ni^{2+} ions per cm^3 (10^{27} per m^3). Reversal of a current in the primary circuit induces an e.m.f., V, across the secondary terminals. Estimate the ratio of the values of V produced if the same current is reversed first at 1 K and then at 4 K.

(For the *free* Ni^{2+} ion, the ground state angular momentum quantum numbers are $S = 1, L = 3, J = 4$.)

Answer

$$V = -\frac{dN}{dt}, \quad N = IL, \quad L = \mu\mu_0 m^2 A\,l$$

where V is the voltage, N the flux passing through the coil, t the time, I the current, L the inductance, m the number of turns per unit length, and A and l are the area and the length of the coil. These formulae can be found in, for example, Bleaney and Bleaney.[2]

$$\mu = 1 + \chi$$

$$\frac{V \text{ at } 1\text{ K}}{V \text{ at } 4\text{ K}} = \frac{\mu \text{ at } 1\text{ K}}{\mu \text{ at } 4\text{ K}} = \frac{1 + ng^2\beta^2\mu_0 J(J+1)/3k_B T \text{ with } T = 1}{1 + ng^2\beta^2\mu_0 J(J+1)/3k_0 T \text{ with } T = 4}$$

where n is the number of paramagnetic ions per unit volume. $n = 10^{27}$, $g = 2, \beta = 9.27 \times 10^{-24}, \mu_0 = 4\pi \times 10^{-7}, k_B = 1.38 \times 10^{-23}$ all in SI units. $J = 4$ for the free ion but nickel belongs to the first transition series and L is quenched when the ion is in a solid. We must substitute S for J in the formula above.

$$\frac{ng^2\beta^2\mu_0 S(S+1)}{3k_B} = 2.09 \times 10^{-2}.$$

The ratio is

$$\frac{1 + 2.09 \times 10^{-2}}{1 + 0.52 \times 10^{-2}} = 1.016.$$

Questions

1) *Manchester, 1971*

The most important contribution to the paramagnetism of copper sulphate comes from the copper ions which have spin ½ and which may be considered to be non-interacting. Show that the magnetization in a field B is given by

$$M = N\mu_B \tanh\left(\frac{\mu_B B}{kT}\right)$$

where N is the number of ions per unit volume and μ_B is the Bohr magneton. Derive the 'high' temperature form of the magnetic specific heat in a constant field B. What is the high-temperature range in this context for $B = 0.5$ tesla (weber m^{-2})? Compare qualitatively the behaviour of the specific heat of copper sulphate in a constant field with that of copper metal in zero field for temperatures up to a few °K.

[Use eqn. 6.2.]

2) *Manchester, 1971*

To what extent is the Weiss model successful in accounting for the temperature dependence of the magnetization of ferromagnetic materials?

The spontaneous magnetization of nickel at room temperature is 4.91 10^5 A m^{-1} and its Curie temperature is 631 K. Assuming $J = ½$ for the magnetic dipoles, use a graphical method to calculate approximately the spontaneous magnetization at absolute zero.

Compare your answer with $N\mu_B = 8.55\ 10^5$ A m^{-1} where N is the number of dipoles m^{-3} and μ_B is one Bohr magneton. Would you expect agreement or disagreement? Give reasons.

[Hint: plot

$$\frac{M_{sat}^T}{M_{sat}^0} = \mathrm{Br}(y)$$

against y (see eqn. 6.3). Then, as $y = Jg\beta B/k_B T = M_{sat}^0 (B_0 + \lambda M_{sat}^T)/nk_B T$, a B_0 of zero gives

$$\frac{M_{sat}^T}{M_{sat}^0} = y\,\frac{nk_B T}{\lambda (M_{sat}^0)^2}.$$

Plot this function against y and see where the two lines cross. See also B. I. Bleaney and B. Bleaney. *Electricity and Magnetism*. Clarendon Press, 1965.]

References

1. K. A. McLauchlan. *Magnetic Resonance.* Clarendon Press, 1972.
2. B. I. Bleaney and B. Bleaney. *Electricity and Magnetism.* Clarendon Press, 1965.

Appendix of Mathematical References

Most of the mathematical knowledge needed to understand this book will have been acquired by the end of the first year at University. General books on 'mathematics for physicists' that might help to fill in any gaps are:

G. Stephenson. *Mathematical Methods for Science Students.* Longmans, 1961.

Henry Margenau and George Murphy. *The Mathematics of Physics and Chemistry* Volume 1. Van Nostrand Company Inc. (2nd edn), 1956.

Morse, Philip M. and Feshbach, Herman. *Methods of Theoretical Physics.* McGraw-Hill/Kogakusha, 1953.

The first of these books is the easiest and is probably adequate for our purposes.

Chapter 1 assumes a slight knowledge of vectors, but not a very great deal and the first chapter of this book should see you through:

D. E. Rutherford. *Vector Methods.* Oliver and Boyd, 1939

It would be helpful to have studied a little statistical mechanics before reading chapter 2. One source of information is:

J. Wilks. *The Third Law of Thermodynamics.* Oxford University Press, 1961 (the first two chapters).

Some wave mechanics is needed for chapter 4 onwards. You should know what a wavefunction is, what a Hamiltonian is, and how to calculate the energy of a state from these two quantities. A very gentle introduction to the whole idea of wave mechanics can be found in:

L. Solymar and D. Walsh. *Lectures on the Electrical Properties of Materials.* Clarendon Press, 1970.

For mathematical treatments try for example:

P. T. Matthews. *Introduction to Quantum Mechanics.* McGraw-Hill, 1963.

R. H. Dicke and J. P. Wittke. *Introduction to Quantum Mechanics.* Addison-Wesley Publishing Company Inc., 1960.

Table of SI Constants

Electron rest mass	m	9.11×10^{-31} kg
Electron charge	e	1.60×10^{-19} C
Speed of light in free space	c	3.00×10^{8} m s^{-1}
Planck's constant	h	6.63×10^{-34} J s
$h/2\pi$	$\hbar$	1.05×10^{-34} J s
Boltzmann's constant	k_B	1.38×10^{-23} J deg^{-1}
Molar gas constant	R	8.31 J mole^{-1} K^{-1}
Avogadro's number	N	6.02×10^{23} mole^{-1}
Bohr magneton	β	9.27×10^{-24} A m^{2}
Permeability of free space	μ_0	$4\pi \times 10^{-7}$ henry m^{-1}
Permittivity of free space	ϵ_0	8.85×10^{-12} farad m^{-1}

Conversion factors

1 electronvolt = 1.60×10^{-19} joule

1 ångström = 10^{-10} metre

1 atomic mass unit = 1.66 10^{-27} kg

Periodic Table of Elements

Period	Group I	II									IB	IIB	III	IV	V	Group VI	VII	O
1	H																	He
2	Li	Be											B	C	N	O	F	Ne
3	Na	Mg											Al	Si	P	S	Cl	Ar
4	K	Ca	Sc	Ti	V	Cr	Mn	Fe	Co	Ni	Cu	Zn	Ga	Ge	As	Se	Br	Kr
5	Rb	Sr	Y	Zr	Nb	Mo	Tc	Ru	Rh	Pd	Ag	Cd	In	Sn	Sb	Te	I	Xe
6	Cs	Ba	La	Hf	Ta	W	Re	Os	Ir	Pt	Au	Hg	Tl	Pb	Bi	Po	At	Rn
7	Fr	Ra	Ac															

Ce	Pr	Nd	Pm	Sm	Eu	Gd	Tb	Dy	Ho	Er	Tm	Yb	Lu
Th	Pa	U	Np	Pu	Am	Cm	Bk	Cf	Es	Fm	Md	No	Lw

Index

Typeset in Great Britain by
Preface Limited, Salisbury, Wilts
and printed in Great Britain by
J. W. Arrowsmith Ltd., Bristol